Label Markets and Applications

A guide to material specifications and regulations for consumer and industrial markets

Other Labels & Labeling books:

ENCYCLOPEDIA OF LABEL TECHNOLOGY
Michael Fairley

THE HISTORY OF LABELS
Michael Fairley and Tony White

DIGITAL LABEL AND PACKAGE PRINTING
Michael Fairley

ENVIRONMENTAL PERFORMANCE AND SUSTAINABLE LABELING
Michael Fairley and Danielle Jerschefske

CONVENTIONAL LABEL PRINTING PROCESSES
John Morton and Robert Shimmin

LABEL DESIGN AND ORIGINATION
John Morton and Robert Shimmin

LABEL DISPENSING AND APPLICATION TECHNOLOGY
Michael Fairley

CODES AND CODING TECHNOLOGY
Michael Fairley

LABEL EMBELLISHMENTS AND SPECIAL APPLICATIONS
John Morton and Robert Shimmin

BRAND PROTECTION, SECURITY LABELING AND PACKAGING
Jeremy Plimmer

DIE-CUTTING AND TOOLING
Michael Fairley

MANAGEMENT INFORMATION SYSTEMS AND WORKFLOW AUTOMATION
Michael Fairley

SHRINK SLEEVE TECHNOLOGY
Michael Fairley and Séamus Lafferty

LABEL MARKETS AND APPLICATIONS
John Penhallow

For the latest list please visit: **www.labelsandlabeling.com**

Label Markets and Applications

A guide to material specifications and regulations for consumer and industrial markets

John Penhallow MA, MBA

Label Markets and Applications

A guide to material specifications and regulations for consumer and industrial markets

First edition published 2017 by:
Tarsus Exhibitions & Publishing Ltd

Printed by CreateSpace, an Amazon.com company.

ISBN 978-1-910507-13-1

Contents

While every care has been taken to ensure the information, charts, diagrams and illustrations in this publication are correct at the time of publishing it is possible that technology, specifications, markets and applications, or terminology may change at any time, or that the editor's or contributor's research or interpretation may not be regarded as the latest accepted guidance in some parts of the world of labels.

The publishers therefore cannot accept responsibility for any errors of interpretation or for any actions, decisions or practices that readers may take based on the publication content and would advise that the latest industry supplier specifications, standards, legislative requirements, performance guidelines, practices and methodology should always be sought before any investment or implementation is made.

Preface

Labels, promoting products of every kind in our shops, are part of our everyday existence. Branded goods rely above all on the label to ensure they are recognized. Recently labels have started to provide a link, via a QR code, to the brand owner's website. But the printed label is not just a marketing tool. It provides health and safety information for foods and medicines, and instructions for use. It also helps to identify parts in almost every industrial process. For many products, especially pharmaceuticals, the distribution chain must be traceable. It is the label, in most cases, that carries this essential track-and-trace information. Security labels help to protect the product, and the brand, from counterfeiters. The label industry is constantly at work to find new ways to beat counterfeiting, fraud and tampering, often using the latest scientific resources.

Labels are used in many different ways. Each industrial or end-user market has its own problems, and its own labeling needs. Each chapter of this book looks at new ways in which labels are being used to increase efficiency in the factory and enhance the marketing of all kinds of retail goods.

This book in the Label Academy series examines each of the main end-user markets in turn. It describes how the label is used to satisfy the consumer, the brand owner and the regulatory authorities in major world markets. It also looks at the changes in label technology, in particular the impact of digital printing, and the ways in which they affect the various end-user sectors.

It provides an information and training resource for all those who are involved in the production, marketing and development of labels, worldwide.

John Penhallow
International Consultant
Label and Narrow Web Industries

About the Label Academy

This book is part of the recommended study material for the Label Academy, a global training and certification program for the label industry. The Label Academy was created by the team behind Labels & Labeling magazine and the Labelexpo series of events.

The Academy consists of a series of self-study modules, combining free access to relevant articles and videos with paid text books (both printed and electronic). Once a student has completed a module, there is an opportunity to take an online test and earn a certificate.

It is expected that a Label Academy qualification will become a standard in the industry – for printers/converters, suppliers, brand owners and designers – and assist in providing a benchmark. In addition to its own training, the Label Academy will aim to become a resource provider to the many existing educational programs in the industry. Accredited training courses will be promoted through the Label Academy website and books will be provided at discounted rates.

The Label Academy concept was pioneered by industry expert Mike Fairley. This was in response to a reduction in the number of dedicated printing colleges and the need to standardize training across the world. The label industry also has its own specific training needs – it has some of the widest range of materials, printing processes and finishing solutions of any printing sector.

We are also working with other training experts and authors to ensure that the Label Academy provides up-to-date and relevant training material for the industry.

The Label Academy is supported by the key trade associations, including FINAT, TLMI and the LMAI.

www.label-academy.com

Label Academy sponsors

Thank you to our founding sponsors, without whom this ambitious project would not have been possible:

Cerm

Cerm designs business automation software solutions to meet the specific demands of flexo and digital narrow web printers. Using the latest technology, our team's focus is on innovation and continuous improvement.

Our automation solutions support each step in the printer's integrated workflow – from estimating to production, shipment and data collection – and provide the feature and functionality printers need to gain efficiency and improve profitability.

Cerm inspires collaboration and helps printers remain competitive in the market and deliver the best products possible. We are proud to sponsor the Label Academy and contribute to the future of the narrow web printing industry.

www.cerm.net

Flint Group Narrow Web

Flint Group Narrow Web has the products, the solutions, and the technical experts to handle any print situation. Providing solutions for food packaging, sustainability, increased bottom line, efficiency, and uptime – delivering the basics needed to run a successful operation, and the expertise to go above and beyond to another level of success.

Our experts provide solutions to your printing problems with the innovative products and services that have made us an industry leader around the world. Wherever you are, we are – available to help you reach your business goals today and into the future.

Continuous improvement is paramount to Flint Group; we are proud to sponsor the Label Academy and the benefits it will bring to the future of our industry.

www.flintgrp.com

Gallus Group

The Gallus Group with its production sites in Switzerland and Germany is a leader in the development, production and sale of narrow-web, reel-fed presses designed for label manufacturers. The machine portfolio is augmented by a broad range of screen printing plates (Gallus Screeny), globally decentralized service operations, and a broad offering of printing accessories and replacement parts. The comprehensive portfolio also includes consulting services provided by label experts in all relevant printing and process engineering tasks. The Gallus Group is a member of the Heidelberg Group and employs around 430 people, of whom 253 are based in Switzerland. The group headquarters is in St.Gallen, Switzerland.

www.gallus-group.com

MPS Systems B.V.

Producing high-quality label printing depends on several factors; one of them is the operator of the press.

As a press machine builder since 1996, MPS Systems B.V. knows how important training and education on subjects like pre-press, label printing and finishing is. For label printers, it is critical that their operators keep up with pre-press and press developments in addition to label trends. Therefore, MPS sponsors the Label Academy, to advance operator's passion for printing, share expertise and help multiply benefits.

The MPS slogans of 'Printers First' and 'Technology with Respect' have always underlined the core philosophy of MPS from press design to operator satisfaction. We develop our presses with a strong focus on user-friendliness and respect for the press operator: Printers First.

www.mps4u.com

HP Indigo

HP Indigo is a global leader in digital printing, with a broad portfolio of digital presses and workflow solutions. Indigo's proprietary Liquid Electrophotography (LEP) technology delivers exceptional print quality for the widest variety of applications including labels, flexible packaging, shrink sleeves and folding cartons. HP Indigo's digital presses match gravure print quality satisfying the most demanding brands.

A division of HP Inc.'s Graphics Solutions Business, Indigo serves customers in more than 122 countries, including many of the top label and packaging converters worldwide.

www.hp.com/go/labelsandpackaging

UPM Raflatac

In a little more than three decades, UPM Raflatac has become one of the world's leading manufacturers of pressure sensitive label materials, developing and leveraging the latest innovations in adhesive technology. Our film and paper label stocks are used for product and information labeling across a wide range of end-uses – from pharmaceuticals and security to food and beverage applications.

We are an engineering driven company with industry-leading products known for their consistent high quality and top performance. We are also known for the high performing supply chain and undisputed leadership in the area of sustainability. UPM Raflatac's dedication to innovation, sustainability and top quality is matched only by our commitment to service excellence. We call it the Raflatouch.

www.upmraflatac.com

About the Author

John Penhallow
International Consultant
Label and Narrow Web Industries

A graduate of Cambridge University and of INSEAD business school, John Penhallow was a Senior Manager with an Austrian packaging company before entering the narrow web industry. He partnered in organizing the first Labelexpo show in Brussels and worked closely to ensure the show's growth.

Over the years, he has worked as a consultant with many of the leading equipment and substrate manufacturers in the label field. He writes reports and articles in French, German and English, and these are published regularly in label magazines both in Europe and in the USA. He also has regular speaking engagements at conferences and seminars.

His company publishes the monthly journal of abstracts 'Label Press International', a bilingual (English/German) publication, which has become a unique information resource for senior managers and deciders in all parts of the label industry.

Chapter 1

An introduction to label markets and applications

Each end-user market is different. Understanding these differences is essential if we are to satisfy the customer, and the brand owner, and work with them to meet present and future needs.

According to a recent report by the market research firm Freedonia Group the forecast growth for global label demand up to the year 2018 is 4.9 percent per year. This means the total value of the global label market will grow to 114 billion USD. For self-adhesives in Europe, Finat's latest report shows that their converter members grew their business, on average, by more than 5 percent in 2016. US and Latin American growth remains strong, and Asian growth is also satisfactory despite the slowdown in China. Wet glue labels still dominate many markets, particularly in emerging countries. In volume terms, self-adhesives account for around one label in three, worldwide. Another growth sector is shrink labels: nearly 10 million sqm of shrink labels were printed in 2014 according to AWA, and annual growth is 5.5 percent.

A recent study from Smithers Pira (The Future of Label Printing to 2022) estimates that the label printing market worldwide will be 35.7 billion USD in 2017, and then grow at 4.4 percent year-on-year to reach 43.8 billion USD in 2022. According to Smithers Pira research, the most widely used labels are web-fed pressure sensitive grades.

FOOD

Food labels account for almost one in three of all self-adhesive labels used worldwide. This amounts to

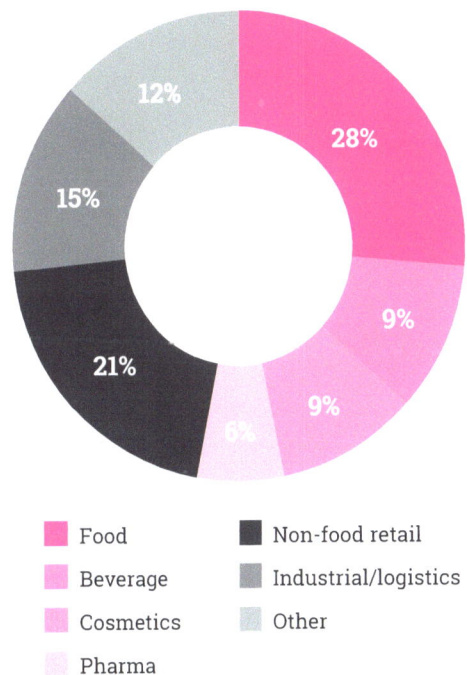

Figure 1.1 World pressure-sensitive label end-use segmentation by volume (sqm). Source: AWA Alexander Watson Associates

some six billion sqm of labelstock every year, or enough to cover the whole of Texas and still have some left over. For wet glue and wraparound food labels annual consumption is estimated at between four and six billion sqm. Foods may be packed in rigid or flexible plastic packs, in cartons or in glass. Each type of pack has its specific labeling imperatives. Temperature is also important: labels for chilled or frozen foods must withstand temperature changes without becoming wrinkled or detached.

COSMETICS, HEALTH AND BEAUTY PRODUCTS

All these products, unlike prescription drugs, are products of choice. And since the consumer can take them or leave them, the attractiveness of the label/packaging is vital if the product is to succeed. Aside from legal regulations, labeling needs to respect both the marketing imperatives (exclusive perfumes are labeled differently from mass-market ones), and the physical ones (bathroom cosmetics must be moisture-resistant, sun creams must resist both water and strong light). Hair coloring, eye makeup and skin care for example each have their specific labeling requirements.

BEVERAGES

The total world market for bottled beverages (both alcoholic and non-alcoholic) is estimated at one thousand billion liters, nearly all of it labeled. Quality regulations are stringent, and ingredients, degree of alcohol and storage recommendations must generally be printed on the label. Alcoholic beverages are frequently taxed, and hence frequently counterfeited. Security closures and hard-to-copy labels offer some protection. Brands are important, as are colors (the red of Coca-Cola labels must be exactly the same world-wide).

Each type of beverage has its own labeling requirements. Waters, soft drinks, wines, beers and spirits must be treated separately. In Chapter 4 of this book we look at the labeling requirements for each of these product groups.

PHARMACEUTICAL AND MEDICAL

In developed markets, and in particular the United

States, the health sector is large and growing. Americans spend an estimated thirteen percent of their GDP on health-related products and services. In almost all parts of the world, regulations and directives for the labeling of medical and pharmaceutical products are very strict. Counterfeiting is rife particularly in developing countries. It has been estimated that one third of all the drugs sold in Africa are fakes. Many of these fakes are useless, others may be actively harmful.

With increasing concern for better protection and counterfeit protection, we can be sure that pharmaceutical label markets (currently estimated at between 1.3 and 1.5 million sqm for self-adhesives) are likely to grow even faster than the pharma market itself. Chapter 5 gives information on the regulatory background for pharma labels, and explains the labeling requirements for prescription and non-prescription drugs. It also looks at the way labels are being used in hospital environments to reduce human error and provide better security for both patients and health-care workers.

NON-FOOD RETAIL AND HOUSEHOLD DURABLE PRODUCTS

Non-food retail products include cleaning and disinfecting products (also called 'under-the-sink'), some of which are hazardous, and all kinds of apparel and shoes. These account for the biggest volumes of labels. Also included in this category are most of the goods you will find in a hardware or do-it-yourself store, or in a sports goods retail outlet. Chapter 6 of this book suggests the optimum type of label (wraparound, wet-glue, self-adhesive, sleeve etc.) for each type of product. It also suggests solutions for labeling 'difficult' products like paints and varnishes.

INDUSTRIAL LABELS

Industrial labels are also a steadily growing end-user sector. These labels are technical, and are generally machine-readable. They are frequently subsumed into a track-and-trace system to provide security and traceability. This is particularly true for car and aircraft parts. Labels for identifying engine parts must adhere and be legible over long periods despite vibration, grease and dust. Tire labels are a classic example of a labelstock developed specifically to adhere to a difficult surface.

Generally, industrial labels require a substrate and adhesive adapted to the product to be labeled. This is explained and expanded on in Chapter 7.

Roll-to-roll printing using conductive inks could pave the way to cheaper electronic components. Although this is not strictly a label application, it is well suited to run on a narrow web press. So far, this is only at the development stage.

SPECIAL LABELS AND CONSTRUCTIONS

There are many special label innovations, with new ones appearing every day. The final chapter of this book looks at just a few of these: re-closable labels, multi-page and 'smart' labels are examples of how ingenious ideas have been turned into profitable products to expand the global label market while helping the end-user solve a problem.

In this chapter (Chapter 8) we also consider the vital question of environmental issues. Labels can and should be recycled but this is often difficult to achieve. For the question of waste liner recycling several initiatives are starting to produce results.

Chapter 2

————

Food labels

————

Food labels have to perform many functions, from product identification and content information, to logistics capabilities. Food labels must also meet the needs of brand owners who require their products to stand out on retailers' shelves. Finally, they must conform to strict regulations (which differ from country to country) and must help public health and safety.

————

MARKET SIZE AND MAIN CHARACTERISTICS

Food labels of all kinds total between ten and twelve billion sqm. However, food labeling and packaging are not just huge markets, they are also slightly different from that of other industries.

Pharmaceutical packaging requires a high level of security; cosmetics packaging seeks to entice the consumer, but food packaging must do both. Food labels and packaging must be sterile to protect the product, as well as to lure the consumer with a promise - such as taste, freshness or 'wellness'. Each category of food label has different requirements: for unpackaged products (fruit, vegetables, sometimes meat) the labeling requirements are not the same as for packed ones. Glass jars (for jams, preserves) and plastic containers use very different label types. For chilled and frozen foods the label needs to resist temperature changes. All these categories, which we look at in more detail later in this chapter, need to have the most suitable label at the least cost.

In every part of the world – but particularly in developed countries – governments and public health bodies regulate all food packaging.

A good starting point for the study of food labels and food packaging is therefore to look at some of the key directives and regulations that food producers and packers must comply with. Designers, as well as label and pack producers must all be aware of these when selecting types faces, substrates, inks, printing processes, label types and methods of application.

DIRECTIVES AND REGULATIONS GOVERNING FOOD LABELING (1)

IN THE EUROPEAN UNION

People everywhere get anxious about the quality of the food they eat. People often worry whether the packaging, or label of what they eat, might contaminate the food. In too many cases where labels/packaging are blamed, these are false alarms. Sometimes there may be contamination but in quantities well below those at which the substances can become a health hazard. The result of this public concern is that laws and regulations control food labeling more and more tightly.

The directives and regulations in force within the European Union and its member states are complex and not always easy to understand. The EU started to harmonize legislation on food contact materials several years ago, but fully harmonized legislation does not yet exist for all materials. Much of the basic regulatory background is contained in Regulation (EC) N° 1935/2004 and in particular article 15 of this regulation. Here is a summary of what this article contains:

- It lays down common rules for packaging materials which come, or may come, into contact with food, either directly or indirectly. It also seeks to protect human health and consumers' interests throughout the European Economic Area (the 28 EU member states plus Iceland and Norway).

- It covers in all 17 different materials, including all papers and boards, plastics, inks, adhesives and coatings. It also covers what it calls 'active and intelligent materials'. Any substances which can reasonably be expected to come into contact or which can transfer their constituents to food are covered by the regulation.
- It determines purity standards and defines what is acceptable for food contact.
- It rules that traceability measures must be in place to make it possible to recall any defective products or provide the public with specific information.
- Where 'active and intelligent materials' (e.g. anti-microbial substances) are used the label must inform about the safe and correct use of the active ingredient.
- Labeling of foods 'shall not mislead the consumer' (many food manufacturers could be in trouble if this regulation were to be interpreted strictly!).

In Europe with its 25 official languages, member states can stipulate the language(s) which must be used on labels. However, they cannot prevent additional languages being used.

DIRECTIVES AND REGULATIONS GOVERNING FOOD LABELING

(2) FDA FOOD LABELING REQUIREMENTS

The U.S. Federal Food, Drug and Cosmetic Act (FFDCA) defines food 'labeling' very broadly. It covers all labels and other written, printed, or graphic matter upon any article or any of its containers or wrappers, or accompanying such article. The term 'accompanying' extends to tags, leaflets, circulars, booklets, brochures, instructions, and even websites.

The Nutrition Labeling and Education Act (NLEA), which amended the FFDCA requires most foods to show specific nutrition and ingredients on the label. Food, beverage and dietary supplement labels that

'In the United States, frozen cheese pizza is regulated by the Food and Drug Administration. Frozen pepperoni pizza, on the other hand, is regulated by the Department of Agriculture. Each sets its own standards with regard to content, labeling and so on, and its own set of regulations which require licenses, compliance certificates and all kinds of other costly paperwork. This kind of madness would not be possible in a small country like Britain. You need the European Union for that'

Source: Notes from a Big country by Bill Bryson

show nutrient content claims (for example 'low fat' or 'contains vitamin XYZ') and certain health messages have to comply with specific legal requirements. Furthermore, the Dietary Supplement Health and Education Act (DSHEA) has amended the FFDCA, in part, by defining 'dietary supplements'. It also adds specific labeling requirements for dietary supplements, and provides for optional labeling statements).

A US government advisory body exists to check compliance of any label with current rules. This service (which must be paid for) is recommended for manufacturers and label/leaflet vendors. Further details can be found at: www.registrarcorp.com/fda-food/labeling/

A recent regulation (May 2016) changes US requirements regarding health criteria to be mentioned on labels. Full details are available at: www.fda.gov/Food/GuidanceRegulation/GuidanceDocumentsRegulatoryInformation/LabelingNutrition/ucm385663.htm

US Manufacturers and those exporting to the US will need to use the new label (see Figure 2.1) by July 26, 2018. However, manufacturers with less than 10 million USD in annual food sales will have an additional year to comply.

FOOD PRODUCTION, DISTRIBUTION AND MARKETING

There are nine billion people on the planet, and rising,

SIDE-BY-SIDE COMPARISON

Original Label

New Label

Nutrition Facts
Serving Size 2/3 cup (55g)
Servings Per Container About 8

Amount Per Serving	
Calories 230	Calories from Fat 72
	% Daily Value*
Total Fat 8g	12%
Saturated Fat 1g	5%
Trans Fat 0g	
Cholesterol 0mg	0%
Sodium 160mg	7%
Total Carbohydrate 37g	12%
Dietary Fiber 4g	16%
Sugars 1g	
Protein 3g	
Vitamin A	10%
Vitamin C	8%
Calcium	20%
Iron	45%

* Percent Daily Values are based on a 2,000 calorie diet.
Your daily value may be higher or lower depending on
your calorie needs.

	Calories:	2,000	2,500
Total Fat	Less than	65g	80g
Sat Fat	Less than	20g	25g
Cholesterol	Less than	300mg	300mg
Sodium	Less than	2,400mg	2,400mg
Total Carbohydrate		300g	375g
Dietary Fiber		25g	30g

Nutrition Facts
8 servings per container
Serving size 2/3 cup (55g)

Amount per serving	
Calories	**230**
	% Daily Value*
Total Fat 8g	10%
Saturated Fat 1g	5%
Trans Fat 0g	
Cholesterol 0mg	0%
Sodium 160mg	7%
Total Carbohydrate 37g	13%
Dietary Fiber 4g	14%
Total Sugars 12g	
Includes 10g Added Sugars	20%
Protein 3g	
Vitamin D 2mcg	10%
Calcium 260mg	20%
Iron 8mg	45%
Potassium 235mg	6%

* The % Daily Value (DV) tells you how much a nutrient in
a serving of food contributes to a daily diet. 2,000 calories
a day is used for general nutrition advice

Note: The images above are meant for illustrative purposes to show how the new Nutrition Facts label might look compared to the old label. Both labels represent fictional products. When the original hypothetical label was developed in 2014 (the image on the left-hand side), added sugars was not yet proposed so the "original" label shows 1g of sugar as an example. The image created for the "new" label (shown on the right-hand side) lists 12g total sugar and 10g added sugar to give an example of how added sugars would be broken out with a % Daily Value.

Figure 2.1 Strict regulations define nutrition labels in USA. Source: FDA (Food & Drug Administration)

and they all need to eat. The packaging and labeling industry is there to ensure that food products get to the consumer in good condition and with relevant information about the product in question.

With a few exceptions, food products rarely go directly from producer to end consumer, except in subsistence economies. Bulk shippers and wholesalers deliver unprocessed or semi-processed food to the people who pack and label it before bringing it to our supermarkets.

THE BRAND OWNERS
Key players for the label sector, the ten biggest brand owners are reckoned to supply two thirds of the pre-

packed food sold in the developed world. The world's leading brand according to retail statistics is Maggi. But of course Maggi is only one of Nestlé's many brands, and Unilever, Nestlé and the other top processed food companies (see Figure 2.2 - Leading Brand Owners) can be said to feed the world. They are by far the biggest users of labels and packaging.

THE DISTRIBUTORS
Distributors, or retailers, are where most people buy their groceries. World leader is Walmart, with around 12,000 outlets in 28 countries, over two million employees and sales just short of five hundred billion USD. In Europe, top retail groups are Carrefour and Tesco, both with sales of over 80 billion USD. For label converters it is important to note the rise in distributors' own brand products (also known as House Brands). For these products the label converter's customer is no longer the brand owner but the retailer. Both brand owners and major distributors wield enormous power over their label and packaging suppliers. Some have been accused of using this power to squeeze prices and impose conditions for payment for special promotions and for (sometimes) punitive rebates. These major customers for the label industry all say that they seek a long-term, mutually beneficial relationship with their label supplier. This is often, but not always, the case.

A QUESTION OF COLOR
Brand owners in particular require their label suppliers to match exactly their brand colors. Any deviation, however slight, can lead to a consignment of labels being rejected. Color conformity is difficult. Modern color management systems have taken a lot of the guesswork out of color matching, but slight differences in the substrate or the ambient temperature can still cause color deviations. The problems are increased when two or more different print technologies are used, for example flexo and digital. A famous case concerns Coca Cola's promotion with hundreds of different first names on the label (discussed in Chapter 3 of this book).

WHERE DO BRAND OWNERS OPERATE?
The developed world is the main market for

COMPANY NAME	ANNUAL SALES ($ MILLION)	EMPLOYEES	MAIN ACTIVITIES
Nestlé (CH)	90,000	340,000	Chocolate, coffee, waters, dairy products, cereals
Unilever (GB/NL)	59,000	170,000	Tea, coffee and wide range of packaged foods
Mondelēz (US)	35,000	107,000	Snacks, candies
Mars (US)	33,000	60,000	chocolates, chewing gum, rice
Kraft Heinz (US)	29,000	46,000	Processed foods, coffee
Danone (F)	25,100	100,000	Dairy products, waters, baby foods
Assoc. British Foods (GB)	21,000	112,000	Teas, cooking oils
General Mills (US)	18,000	43,000	Wheat-based products
Kellogg's (US)	13,000	32,000	Cereals, snacks, cookies

Figure 2.2 Leading brand owners. Source: Cab Penhallow Consultancy

processed packaged foods. Consumption per head is highest in North America, followed by Japan and Western and Central Europe. Asian markets, and particularly China, are expanding fast, but from a very low base. India, which is home to fifteen percent of the world's population, has little in the way of modern distribution networks. Buying patterns in India revolve around local stores, and laws to protect small shopkeepers prevent the growth of supermarkets. For Africa (outside major cities in a few countries like South Africa, Kenya and Ivory Coast) packaged fresh foods are hard to find.

BRAND PROTECTION

Counterfeit food products are a major problem in the developing world. Even in developed countries consumers frequently buy by glancing at the brand name and the color, without looking for subtle differences (e.g. Hienz ketchup can pass for Heinz ketchup). This is one of the reasons why brand owners are so concerned about color consistency. The problem is that most foods, unlike luxury baggage or silk scarves, have a low unit value. This rules out most forms of anti-counterfeit protection. QR codes are sometimes used, but for low-value goods they are more marketing tools than guarantees against forgeries. Only high-value foods like foie gras, caviar or truffles can be economically protected by hard-to-copy features like holograms and intricately colored labels. Tamper-evident closures are a different matter. They take many forms, but one of the most popular is the tamper-evident label or seal.

Figure 2.3 shows examples of tamper-evident protection. This reassures the law-abiding consumer

Figure 2.4 Poor print quality can expose the counterfeit product

Figure 2.3 Tamper-evident closures

and discourages the cheats who empty food packs and refill them with diluted or inferior product. In some cases, label films are heat-sealed to the container to form a coating which protects against tampering and helps to keep the product fresh.

Another technique (Figure 2.4) uses inks to detect differences between fake and genuine foods. Further information on beating the counterfeiter with overt and covert security technologies can be found in Chapter 4 ('the battle against counterfeiters').

Figure 2.5 Specialty short-run food labels, printed here on a Primera digital press

DEMOGRAPHICS AND CULTURES – NEW FOODS, NEW CHALLENGES MEAN MORE LABELS

Everyone knows that in developed countries the average age of the population is rising. What does that have to do with labeling? Answer, quite a lot, because older people buy smaller packs. The same is true for many single-person households. Smaller packs mean more labels. In the European Union, with its 24 official languages, regulations require foods to be labeled in the language of the country or region in which they are to be sold. We see also the trend by brand owners to target specific segments of the population. Targeting

teenagers, millennials or religious minorities means more and shorter run lengths for the label converter. In Europe and in North America there are also many different variations of food products – ethnic (Indian, Chinese, Jewish, for example), low fat/healthy eating, gluten-free, low sugar, and many others.

All this leads to multiple label variations and shorter runs. This trend, particularly in Europe, gives rise to greater use of digital label printing for foods (see Figure 2.5).

The trend of smaller packs is not limited to the developed world. For example, in India some packaged foods are often sold in single-use sachets to attract consumers with limited budgets. Cultural factors also

CATEGORY	PRINCIPAL DECORATION TECHNOLOGIES
Unpackaged foods	Paper/filmic self-adhesives
Canned goods	Wet-glue, wraparound, direct printing
Glass jars/bottles	Wet-glue, wraparound, leaflet labels
Squeezable plastic bottles	Stretchable filmic self-adhesives, sleeve labels
Rigid plastic containers	IMLs, wraparounds, self-adhesives
Chilled goods	Moisture-resistant self-adhesives, direct printing
Frozen goods	Temperature-resistant filmic self-adhesives
Price-Weigh	Direct thermal adhesive labels

Figure 2.6 Principal decoration technologies. Source: Cab Penhallow Consultancy

affect how a product's label is presented. Cows are used on many labels in Europe and America; they would not be acceptable in India. In China, where yoghurt is seen as both a gift and a health product, the top two fresh yoghurt brands with yoghurt drinks are proving increasingly popular as a way to aid both social advancement and digestion.

One constant feature, at least in the developed world, is the general rise in packaged, convenience foods. A report by analysts Smithers Pira in 2014 predicts a three percent annual rise for European packaged food and drink consumption to 953 billion packages by 2020. With this trend will come new demands on packaging and labeling for ready meals, prepared chilled foods and on-the-go snacks.

PRINT TECHNOLOGIES AND LABEL TYPES FOR FOOD LABELING

All conventional and digital print processes are used for food labels. For primary labels, which go on the product itself, the appearance of the label is very important. Many studies have shown that shoppers in supermarkets are attracted to the label first, before looking at the brand, the ingredients or the price. This means that the design, the colors and the shape of the label are vital to its success in selling the product.

A walk round any supermarket will show that, although self-adhesives dominate, all label types are used in food labeling. Figure 2.6 summarizes the main categories.

Where the label is **applied directly** to a fruit or vegetable, food-grade inks and adhesives are essential as the risk of migration is high. For fresh meat the same applies, with the additional requirement that barcodes and text must remain legible as the label ensures traceability from the butcher's shop back to the farmyard. Moisture-resistant adhesives are also needed as these goods are generally chilled, often for long periods.

Canned goods are for the most part low-value items (canned foie gras is among the exceptions), and they mostly have wet-glue or wraparound labels, or are direct-printed. Wet-glue is also used frequently for food products in glass jars (jams, preserves, spices etc.), but

Figure 2.7 Squeezable containers are not easy to label

Figure 2.8 Tubular casings. Source: KPG

self-adhesives are increasingly used. A jar will usually have a prime label on the front and a second label (often in several languages) on the back, giving ingredients and legally required information. The top of the jar sometimes has a multi-layer leaflet labels, with suggestions for use, or with multiple languages.

Plastic bottles may be rigid or, more usually, squeezable, for example, bottles for ketchup, mayonnaise and mustard (see Figure 2.7). This surface presents a particular problem. To prevent the label from creasing or peeling, a self-adhesive labelstock with a conformable face material is needed. Traditionally, conformable labels have been made from polyethylene. PE is sufficiently deformable to make excellent squeezable labels, but it has a tendency to become hazy with exposure to light or humidity. Recently, other squeezables have been developed; made from biaxially oriented polypropylene (BOPP). Polyolefins are also a popular facestock for squeezable labels. Occasionally, plastic bottles may be decorated using sleeve labels.

Rigid plastic containers for food products (yogurts are a typical example) are decorated using a wide variety of label types. Wraparounds, wet glue and self-adhesives can all be found. Many products of this type are chilled, and for these synthetic label materials are mainly used. High-volume products like dairy spreads will often be decorated with in-mold labeling (IML). IML was initially carried out by blow molding, and this technology still dominates in North America. In Europe developments using injection

molding or thermoforming with reel-fed systems have increased the speed and efficiency of the labeling process. The original concept involves coating the reverse side of the label with a heat seal layer, followed by a substrate material printed with heat resistant ink, and finally a heat resistant varnish coating. IML labels are most often made of polypropylene. After being printed they are often cut and stacked ready to be applied, though roll-to-roll application is proving popular. Historically, IML has only been cost effective for large volumes, but modern print and application technology has reduced the volume threshold. Consultancy group AWA reckons that by 2017, world consumption of IMLs will top one billion sqm, more than half of it in Europe. The main use of IML is for packing products like butter, margarine, ice creams or edible fats. It is also extensively used for labeling motor oils and detergents. IML packaging has the advantage of being easy to recycle.

Artificial tubular casings are not strictly labels, but are frequently printed and converted on conventional label presses with flexo, letterpress or combined printing. They are used particularly in Central and Eastern Europe (see Figure 2.8).

Applications include smoked fish, pet foods, and of course a huge range of sausages. Casings, which are sometimes printed on both sides, are printed flat, although the web is actually an extruded tube without a join. After printing, the tubular casing is filled under pressure and sterilized at high temperature in an autoclave. The inks used must withstand sterilizing, as well as rough handling during transport and retailing.

Labeling of frozen foods must obviously be done with materials which can withstand temperatures below -20C without falling off or becoming wrinkled. An added problem for the label converter is that some products may be labeled when hot, others only when deep-frozen. Adhesives and label materials must be adapted to whichever processing sequences are used.

In terms of numbers (though not of value) price-weigh labels are possibly the most universal food product label after prime labels. These labels are generally direct-thermally printed. Traditionally, the substrate used for price-weigh labels has contained bisphenol A (BPA). This chemical exhibits estrogen mimicking, hormone-like properties that raise concern about its suitability for contact with foods. Both the American FDA and the European Food Safety Authority have expressed concern and have banned the use of BPA in certain products, for example, bottles for babies. In February 2016, France announced that it intended to propose BPA as a 'Substance of Very High Concern' under RFACH regulations.

APPLICATION OF FOOD LABELS

Applicators for wet-glue labels are used for certain high-volume labeling processes. The equipment is expensive and very often, when it wears out, it is replaced by self-adhesive labelers. These vary from small, semi-manual dispensers to high-speed fully automatic lines able to handle 1,000 products each minute. A more detailed description of food label applicators can be found in the Label Academy Label Dispensing and Application Technology book.

MIGRATION

For Europe, regulation (EC) No 1935/20041 requires that materials and articles which, in their finished state, are intended to be brought into contact with foodstuffs, must not transfer any components to the packed foodstuff in quantities which could endanger human health, or bring about an unacceptable change in the composition or deterioration in organoleptic properties. Over and above these legal requirements there is a groundswell of opinion, particularly in Germany and Scandinavia, that packaging of all kinds is ethically dubious and ecologically hazardous. It is therefore particularly important for the label industry as a whole to develop and promote 'safe' food labels which can be easily recycled.

MIGRATION (1) INKS

The risk that chemicals may leach through labels and packaging is real. The risk that this might harm people's health is mostly imaginary. The fear arises through most people's ignorance of what chemicals there are in ink, and these fears are fanned by the popular press which loves to blame manufacturers for real or imagined contamination. As a result, label converters are turning preemptively to more expensive low-migration inks. What is needed to ensure low or zero migration of ink into a food product? The answer is more complicated than the question, because it depends not only on the ink, but also on the substrate, the drying/curing process and even the shelf life of the food in question. Some ink manufacturers are promoting UV LED inks not only for the energy economy but also because this drying technology can be better controlled than traditional mercury lamp drying. Most ink manufacturers now guarantee their low migration label inks, but these guarantees are hedged with exceptions like 'when correctly used'. Label converters should dialog with their ink suppliers to find the right compromise between security and cost.

MIGRATION (2) SUBSTRATES

In 2015, Aldi, a major discount retailer based in Germany, sent a circular to all its label and packaging suppliers about the risk of contamination through mineral oils. This letter included the following warning: 'Aldi Süd's goal is that in the house brands of its food sector, mineral oil components can no longer be found in foods. For this reason, we are asking you to take measures to ensure compliance with this requirement'. Aldi's defensive stance is not unconnected to the Great Advent Calendar Controversy; an environmental group claimed that the paper used for advent calendars contained recycled fibers contaminated with ink residues. In fact, the residues were only measurable in two of the twenty or more makes of calendar, but the damage was done. The German Association for Food Regulation came back with the argument that 'traces of mineral oil are impossible to avoid in many raw materials due to the processes used, can easily result in wrong conclusions and unjustified reactions'. For label converters there is less risk. Paper face materials are mostly made from virgin fibers, and synthetic substrates have (so far) been less often accused, and are available in 'high barrier' options from several producers. UPM Raflatac for example has developed multi-purpose top-coated films designed for labeling applications on rigid packaging where good water, chemical, and oil resistance is required. It also offers a range of ultra-thin face materials making it possible to create innovative and intricate food label designs on rigid plastic and glass.

MIGRATION (3) ADHESIVES

Adhesives used for self-adhesive labels can also be a potential source of migration and contamination, particularly when these labels are applied directly to foodstuffs. The two most commonly used adhesives for pressure-sensitive labels are hotmelt and acrylic emulsion. Nearly all adhesives stick because they contain resins, and the lower the resin content, the lower the initial tack. This problem is even more acute with filmic labels. However recent developments using a multi-layer adhesive technology have resulted in a virtually resin-free adhesive, which can be used for filmic food labels as well as for moist or fatty surfaces.

ADVANCED LABEL DESIGNS AND TECHNOLOGIES IN THE FOOD SECTOR

Advanced technology applications for food labeling include extending the shelf life of packaged foods. This reduces food waste, and is mainly done by using sealed containers with nitrogen gas flush or partial vacuum. Labels come into play when in direct contact with the product (for some foods the label is used to cover completely and seal a rigid plastic container).

Time for tea

QR codes which provide (for example) cooking recipes when tapped with a smartphone are becoming common. A more original use of these codes comes from a distributor of teabags. A QR code on the back of each pack is linked to a piece of music that not only sets the mood, but also acts a timer – when the music is over the tea is ready to pour out!

The use of moisture, oxygen, ethylene, and CO_2 sachet-type absorbers are all being used in combination with labels to extend shelf life. Recent experiments have also involved the use of zinc oxide nanoparticles, which have an antimicrobial effect.

Time-temperature indicators are not new, although accurate and reliable ones are too expensive to be used on the majority of packed foods. A more effective way to ensure the freshness of foods is to control the chill or cold chain. This however can be done more effectively by temperature indicators placed on pallets rather than on individual products.

The future may be another country, but one thing is as certain as anything can be: the rapid growth of pre-packaged/labeled foods, particularly in emerging countries, means that food labels will continue to be a vital end-user sector for the majority of label converters world-wide.

POINTS TO REMEMBER FOR FOOD LABELS
- About one third of all self-adhesive labels are for food
- Strict regulations and laws govern the position, size and content of food labels
- The label is an essential selling tool
- Migration of substances from the label into the product has become a major public health issue
- A small number of brand owners and retailers dominate the market for packaged foods
- Innovative labels are being used to enhance products or prolong their shelf life.

Chapter 3

Labels for beverages (wines, spirits and non-alcoholic drinks)

It is logical to group food and drink together. However, for the label converter, beverage labeling has many distinctive features, which is why we are treating it separately here. Also, Finat has designated beverage labeling as the end-user sector with the most potential for self-adhesives. At present it accounts for 8-9 percent of the total self-adhesive market.

GLOBAL MARKET SIZE AND MAIN CHARACTERISTICS

The total world market by volume of all bottled beverages, including waters, wines, beers and spirits is of the order of one thousand billion liters, or 140 liters for every man, woman or child in the world. Bottled waters are in volume terms the biggest category, accounting for some 15% of the total, followed by non-alcoholic drinks. Waters and soft drinks are also the fastest-growing sectors, with rapidly expanding sales particularly in Middle East countries. The worldwide beer market is stagnant, but with some rapid growth areas like Africa, where branded beers are replacing home-brews. So-called 'craft' beers – low-volume beers, generally from small breweries – are an expanding business particularly in the USA. Wine production fluctuates with weather conditions. Over the past decade South American, Australian and New Zealand wines have taken an increasing share of world markets. However, half of all the world's wines come from just three countries: Italy, France and Spain. Spirits consumption worldwide is growing. The important Chinese market has stabilized but still has good growth potential.

Before looking in more detail at the label business in each of these subsectors, it is useful to review the most important directives and regulations governing print content, as well as substrate and application methods, for the various beverage labels.

REGULATORY BACKGROUND – ALCOHOLIC BEVERAGES – EUROPEAN UNION

For drinks containing alcohol, labeling regulations everywhere require an indication of the alcohol content, and in most cases also the country of origin and a health warning and storage instructions. For EU countries there are standard regulations (see ref 1169/2011 for 'information which must be provided' on the label). As regards public health warnings, the European parliament in April 2015 voted a resolution 'Calling on the Commission to begin work immediately on the new EU Alcohol Strategy (2016-2022) with the same objectives, updating the regulatory framework so as to assist national governments in dealing with alcohol-related harm, to support monitoring and the collection of reliable data, to encourage prevention and health promotion and education, early diagnosis, improved access to treatment, continuous support to those affected and their families, including counseling programs, to reduce traffic accidents caused by drink-driving and to better differentiate within drinking patterns, behaviors and attitudes towards alcohol consumption'. This resolution has not yet been adopted by the Commission, but many suppliers of alcoholic drinks are already adding extra health-care warnings, for example, those aimed at women during pregnancy.

An extract from EU labeling requirements:

Alcoholic beverages having alcohol by volume of 1.2% and above are exempt from ingredient list and nutrition declaration requirements (Art. 16, Res. 2015/2543RSP). Member States may... maintain national measures as regards the listing of ingredients.

Allergens: Sulphur dioxide and sulphites at concentrations of more than 10 mg/kg or 10 mg/liter in terms of the total SO2; exempt from labeling requirements are cereals containing gluten and nuts when used for making distillates and fish gelatin or isinglass used in wine fining.

Alcoholic beverages with added liquorice at concentrations of 300 mg/l or above: 'contains liquorice – people suffering from hypertension should avoid excessive consumption' shall be added immediately after the list of ingredients.

Even within the EU some countries have specific extra labeling requirements: Austrian wine labels for example must show the origin of the wine and the amount of sugar and alcohol indicated as dry, semi-dry, sweet, or semi-sweet. The United Kingdom stipulates that the wording on all labels shall be 'in British English'.

REGULATORY BACKGROUND – ALCOHOLIC BEVERAGES – UNITED STATES

These are laid down principally by the US Tax and Trade Bureau (TTB) which in addition to the usual stipulations on alcohol content and country of origin also defines the conditions to be met before a beverage can be called 'low carbohydrate'. For each type of beverage (wines, beers, spirits) the TTB provides detailed rulings on label content, font size and many other requirements .

REGULATORY REQUIREMENTS – NON-ALCOHOLIC BEVERAGES

Bottled waters are a huge market (ten billion USD in the US alone) and are subject to strict labeling requirements particularly regarding their origin and the quantities of minerals they contain. Certain waters are used in the treatment of illnesses, and incorrect or inadequate labeling information can have serious consequences. For waters marketed in the EU, Directive 2009/54/EC regulates the marketing and exploitation of natural mineral waters. Certain provisions of this Directive are also applicable to spring waters. Commission Directive 2003/40/EC establishes concentration limits and labeling requirements for natural mineral waters and the conditions for using ozone-enriched air for the treatment of natural mineral waters and spring waters.

Natural mineral waters are subject to an authorization procedure carried out by the competent authorities of the EU Member States.

For waters in Europe (and to some extent in the rest of the world) brand names and places of origin are important marketing tools. The label being an essential part of the marketing mix must conform exactly in color with the brand owner's requirements. Typical example of these strongly branded waters are: Perrier, Volvic, and San Pellegrino.

For the United States the FDA gives only succinct labeling guidelines, namely that the label must include the bottler's company name and contact details, plus 'some quality details'.

For other non-alcoholic drinks like colas and juices, labels generally (in most countries) must show the percentage of fruit, if any, and the sugar content.

PRODUCTION, DISTRIBUTION AND MARKET STRUCTURE

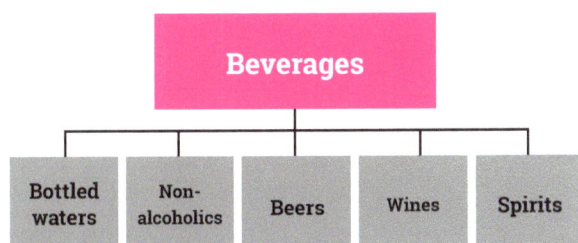

```
                    ┌──────────────┐
                    │  Beverages   │
                    └──────────────┘
     ┌──────────┬──────────┬──────────┬──────────┐
┌─────────┐┌─────────┐┌────────┐┌────────┐┌────────┐
│ Bottled ││  Non-   ││ Beers  ││ Wines  ││Spirits │
│ waters  ││alcoholics││       ││        ││        │
└─────────┘└─────────┘└────────┘└────────┘└────────┘
```

Market structures are very different for each type of beverage.

For bottled waters, Nestlé with bottled water sales of 30 billion USD worldwide is among the leaders (it owns Perrier, San Pellegrino and other well-known brands).

For colas and other non-alcoholic drinks Coca-Cola is market leader with annual sales worldwide of 50 billion USD, followed by Pepsi with 35 billion USD. Other players in these markets are considerably smaller.

The beer market is also increasingly concentrated, dominated by Benelux-based Heineken and InBev (2016 sales 22 billion and 44 billion USD respectively) and South Africa's SAB Miller. InBev's 100 billion USD acquisition of SAB Miller will create a near monopoly in certain markets (see Figure 3.1). Total world beer sales at retail prices are estimated at between 600 and 650 billion USD.

The wine market by way of contrast is very fragmented. Total world production is estimated at 276 million hectoliters, and over half of this is produced in just four countries, France (17 percent), Italy (16 percent), Spain (14 percent) and USA (11 percent). Long-term production of wine is falling as growers switch from low-margin table wines to classier vintages. Significant is the rise of Australian and New World wines which over the past decade have increased volumes substantially. The other joker in the pack is China. Statistics on its wine production are not necessarily reliable, but according to some observers the rate at which new vineyards are being planted in China could make that country the largest wine producer in the world in the longer term as the

- **Heineken**
- **SABMiller**
- **InBev Anheuser-Busch**
- **Carlsberg**
- **United Breweries**
- **Diageo**
- **Tsingtao Brewery**
- **Molson Coors Brewing Company**
- **Boston Beer Company**
- **Beijing Yanjing Brewery**

Figure 3.1 World's leading brewers in order of size. Source: Cab Penhallow Consultancy

Wine production by country 2015

	Units: million hectolitres	Change 2014/15
Italy	48,9	+10%
France	47,4	+1%
Spain	36,6	-4%
USA	22,0	+1%
Argentine	13,4	-12%
Chile	12,0	+23%
Australia	12,9	-20%
South Africa	11,3	0%
China (2014)	11,2	0%
Germany	8,8	-4%
Rest of world	51,2	
Total	**275,7**	**+2%**

Figure 3.2 Wine production by country 2015. Source: O.I.V.

new vineyards reach maturity. Already, China (including Hong Kong) is reckoned to be the largest red wine-consuming nation in the world. Its imports come mainly from France, followed by Australia and Spain. Consumption of wines has increased rapidly in recent years and per capita US consumption now stands at 12 liters per year, a rapid increase, but still a long way behind France and Italy (44 liters and 39 liters respectively).

The spirits market is different again, with two companies, Bacardi and Diageo, dominating. Between them they control two-thirds of the world spirits market. Well-known brands of whisk(e)y, vodka and gin are among Diageo's brands (it also owns the famous Guiness stout). This UK-based company had 2015 sales of 10.8 billion GBP (16 billion USD). Bacardi, with headquarters in Bermuda, has worldwide sales of 4.5 billion USD. The fastest growth in spirits consumption is in the developing world and in particular in China. Chinese demand for imported spirits (especially whisky and brandy) is booming and some industry experts believe China could soon be the largest spirits market in the world.

Self-adhesive beverage labels currently account for some two billion sqm of labelstock. The total volume of other types of beverage label (mainly wet-glue) is between three and four billion sqm, which gives self-adhesives good prospects for growth.

TYPES OF LABEL MOST USED FOR BEVERAGES (1) – BOTTLED WATERS

Often neglected as the 'poor relative', labels for waters are a vast market. With each person in the world consuming over twenty liters per year, that works out at between 120 and 150 billion labels every year. By far the dominant labeling technology is wraparound, with labels printed in one or two colors. Print runs are long, and gravure is the most used print technology, followed by flexo. Most waters come in plastic bottles and recyclability (see below) is a major concern for public bodies, bottlers and label converters. Premium brands of water will often have two or more labels. One or two of the world-famous premium brands of water use glass bottles. Glass is also used for water distributed to HBR (hotels, bars and restaurants) and labels for these must adhere

Figure 3.3 Coca-Cola labels run by Eshuis on an HP Indigo digital press

securely until the empty bottle is washed, and must then be easily detached – not an easy task.

TYPES OF LABEL MOST USED FOR BEVERAGES (2) – NON-ALCOHOLIC DRINKS

Carbonated soft drinks and fruit juices are the main categories in this sector. Market leader Coca Cola uses almost exclusively wet glue labeling. However, when the company was preparing its 'Share a Coke' campaign (with different first names on the label) in 2013 a problem arose. The company has always ordered its labels in very large quantities. Suddenly it needed to order just a few thousand of every label type – and there were thousands of different types. A self-adhesive label converter in the Netherlands rose to the challenge. He 'recruited' half a dozen converters in different parts of Europe, all using the same model of HP Indigo press, and together they delivered what was called 'the most complex label order ever seen'. Apart from the logistics, the biggest problem was ensuring color conformity: every label had to be in exactly the right shade of Coca Cola red. Since this first large-scale experiment, several other beverage manufacturers have copied the idea, which is made possible by today's digital label presses. This has helped the self-adhesive label to broaden its appeal to beverage manufacturers, many of whom still use simple wraparound labels.

Figure 3.4 When tapped with a smartphone app, the characters on the bottles come to life

Figure 3.5 Direct printing onto plastic containers

The juice market is another big beverage sector where branding is important. Self-adhesive labels are widely used to make the product stand out on the supermarket shelf. Shrink sleeve labels are also increasingly used, and they have several advantages. Their 360° coverage of the surface makes them attractive for designers and brand owners, and by using opaque films the shrink label is light resistant, an important criterion for perishable liquids. They can be designed to give full or partial coverage of a bottle, and can also provide tamper-evidence. Being printed on the inside makes them scratch-resistant. Films used are PVC, PETG, PLA and EPS Foam, each type having different shrinkability. Designing shrink sleeve labels is difficult, requiring 3D programs like Esko Studio, Pigmentz or Adobe Illustrator. Special programs exist to pre-distort the designs. All print technologies can be used, but today's shorter runs favor flexo or digital. Steam, infrared or hot air are used to apply the sleeve to the bottle.

TYPES OF LABEL MOST USED FOR BEVERAGES (3) – BEERS

Unlike wines, which are mainly in glass bottles, beers are packed in three main ways – bulk (draught), can and bottle. Worldwide, the bottles account for one third of total beer consumption. Labels for mass-market beers are still mostly wet-glue. However one major brewer, Heineken, is developing what it calls 'brand experience', and is using self-adhesive labels as an integral part of it marketing plan. In this case the labels are clear-on-clear to give a no-label look to the bottle. Several other brewers have copied this idea. Another labeling innovation, popular in the United States where beer is often drunk from the bottle, is to give a 'tactile feel' to the label. This is done by embossing, screen-printing or (more recently) by digital printing.

Another development in beer markets is the growth of so-called 'craft' beers. Originally home-brewed by lovers of 'authentic' brews, they are now a rapidly growing part of the market, particularly in the United States. According to the Brewers' Association, there are today more than 3,500 craft breweries in US alone, they sold 26 million hectoliters in 2014, and had sales of 20 billion USD. The trend is well established in England where it arose out of the 'real ale' campaigns of 20 years ago, and even in Germany 12 percent of all beer sold in 2014 was craft beer (according to Mintel Market Research). The importance of the craft beer revolution is that these beverages are mostly sold in bottles, with self-adhesive labels, and with many varieties. This is helping label converters to gain ground in the beer sector, so long as they can offer premium printing quality in a wider variety of substrates than is normal with the Heinekens of this world. Label converters who have invested in new technologies such as digital printing capacity are able to meet the short-run requirements of the craft beer sector, which includes special effects such as tactile varnishes, hot/cold foils, embossing and inks which change color when cooled.

Because craft beers are more expensive, labels are sometimes provided with NFC tags; these are thin, flexible labels that integrate with a product's packaging or label and can be read with the tap of an NFC-enabled smartphone, enabling the consumer to dialog with the brewer or distributor. For the self-adhesive label business there is however a downside to technological innovation in beer labeling: the arrival – or rather the return after several decades of absence – of direct printing. The equipment manufacturer KHS has developed a digital direct print engine configured with print heads from Xaar and using low-migration inks from Agfa. The manufacturer claims a resolution of 1080 x 1080 dpi and a speed of up to 36,000 PET bottles per hour. The world's first 'Direct Print Powered by KHS' system in industrial production with direct printing on PET beer bottles was installed in the Martens Brouwerij (Belgium) in mid-2015. Another contender is Krones with its Direct Print System DecoType for decorating plastic containers, said to run at up to 12,000 containers/hour with high print quality (see Figure 3.5). According to Krones this can enable digital container decoration to print onto shapes where labels cannot be used.

It also avoids overproduction of labels, needs no adhesive and so reduces storage costs.

This development of direct printing could have repercussions far outside beer labeling and could inhibit the growth of self-adhesive technologies over the whole beverage sector.

TYPES OF LABEL MOST USED FOR BEVERAGES (4) – WINES

With very few exceptions, wines are sold in glass bottles, and the only two labels types used are wet-glue and self-adhesive. Traditionally, wet-glue labeling has dominated in Europe. For low-value, high-volume wines it still dominates. Wet-glue bottle-labeling units are expensive, and they are durable. Bottlers tend to wait until the last possible moment before considering a change to self-adhesive labeling. The self-adhesive

Figure 3.6 UPM Raflatac has developed label materials specially for white wines

wine label offers many advantages: any shape of label is possible, and two or even three labels can be accurately applied in a single pass. The disadvantages are cost, and speed of applications. The latest generation of fully automatic pressure-sensitive labeling lines run at speeds close to those of wet-glue applicators, but total applied cost is still higher for self-adhesives. However, the wine market worldwide is changing. The market for cheap table wines is declining, and growers are trading up to higher quality wines. For labeling, run lengths are getting shorter, so the fast set-up time of self-adhesive labelers is a big advantage. Finally, many smaller wine growers who in the past would have sent their wine in bulk to be bottled and labeled now find that they can do the job themselves with only modest capital investment in a semi-automatic self-adhesive labeler. Particularly in Europe, wine label converters can be found clustered thickly around the continent's main wine-growing areas. Although flexo printing dominates for these converters, many use venerable letterpress machines. In fact, a wander round some of these small label converters' shop floors can be as good as a trip to the museum. Digital label printing is making inroads into the wine label sector as growers, cooperatives and shippers call for special labels for supermarkets, wine stores and hotel/restaurants – and they don't want to hold stocks of labels which they may never need. For these reasons, wine label converters are 'going digital' to provide their customers with just-in-time, short-run labels.

Smart labels for Burgundy

Belgian authentication specialist Selinko in partnership with NXP Semiconductors now supplies smart labels for wines made on the Geantet-Pansiot estate in Gevrey-Chambertin. The new microchip-enabled wine labels will enable the estate to safeguard its distribution network in order to protect its wines from counterfeiting, combat gray markets and also to communicate with its customers directly and in a targeted manner. The customers, for their part, can authenticate their bottle and obtain information about the wine (vintage, production, serving temperature, storage) and the estate, simply by tapping the back label with their NFC smartphone. Similar smart devices are already used on Grand Cru wines. This is the first time smart labels are being used for more modestly prices vintages.

When launching a brand of Cachaça (a kind of rum) on the UK market, the distiller wanted a label that would stand out. The design chosen used Tintoretto Crystal Salt material for a rich tactile feel to reflect the Salto sugarcane green, with subtle sepia imagery. Added embellishments included a reflective gold foil to give the product a premium look. The fine foil detail on this label, printed and finished by AJS Label on HP Indigo and Newfoil equipment, is exceptional and typical of the attention to detail given to spirits labels.

Different wines need different substrates, and all the major labelstock producers now have tens, even hundreds of grades specially developed for the wine sector (see Figure 3.6). Labels for white wines, rosés and champagnes are generally coated, wet strength papers, since they must withstand chilling and immersion in iced water without wrinkling or falling off; mass-market rosés sometimes carry a clear-on-clear label, but this is the exception rather than the rule. For red wines the fashion is to have a textured label, either with a paper face material or a white filmic substrate made to look like paper. For all types of wine except the most expensive, gold foiling is used extensively to make the bottle stand out on the shelf. Several label converters have developed applications which enable their customers to design their own wine labels from a set of patterns; interestingly, the customers are free to put whatever text they want on the back label (which contains the legally required

information). 'That way', one label converter explained 'If they make a mistake, it's not our fault'. The back (or contra) label for wines is in nearly all cases a white label printed in one color. The same conditions (moisture resistance) apply as for the front label.

The wine label does not normally need to be recyclable, since in most countries, Germany is an exception, glass wine bottles are not washed and re-used. Glass being a total barrier, there is no question of unwanted migration either.

Self-adhesive and wet-glue labels account for over 90 percent of the wine labeling sector worldwide. The remaining ten percent is mostly bag-in-box packing, increasingly popular for cheaper wines. There is also direct printing, made possible through recent technical advances, and sleeve labeling. Both of these are at present very small niche markets.

TYPES OF LABEL MOST USED FOR BEVERAGES (5) – SPIRITS

In most of the world you are allowed to brew beer in your bath and wine in your washtub, but you may not distil spirits. This is partly for public health reasons (wrongly distilled spirits poison you) but mostly since spirits are heavily taxed. With just two companies controlling two-thirds of the business, the global spirits market is highly concentrated. It is also highly regulated. Some governments require distillers to seal

Figure 3.7 AJS, HP Indigo and Newfoil contributed to this remarkable spirits label. Source: AJS Labels

Figure 3.8 Game-changing smart technology for spirits labels. Source: L&L

including screen printing, embossing and foiling. Different overt and covert security features are used. While contra-labels for wines are mostly white, in the case of spirits the contra label is colored and complex, often giving cocktail suggestions as well as health warnings.

The proliferation of brands and sub-brands of spirits makes for shorter runs. This together with distillers' requirements for just-in-time deliveries means that spirits labels are a market opportunity for the self-adhesive label converter, particular if equipped with a digital press. For top-of-the-range spirits, NFC-based smart labels are being used. Constantia Flexibles in partnership with Thinfilm Electronics has developed smart labels which are uniquely identifiable and not only provide anti-counterfeit protection but also offer the brand a two-way communication with the consumer (see Figure 3.8). At present these labels are cumbersome and expensive, such that they can only concern expensive items, but the technology is likely to become cheaper as it grows and spreads.

RECYCLABILITY AND THE ENVIRONMENT
Plastic bottles are increasingly used for non-alcoholic drinks, and occasionally for beer. In many cases the plastic is PET. PET bottles are generally not washed and re-used; they are recycled by shredding, and are

each bottle with a sequentially numbered self-adhesive security tape or tax stamp. There is a small but flourishing black market in 'genuine' spirits labels which are sold to crooked distillers. Label converters need to be equipped to count spirits labels with great care and to ensure that surplus or slightly defective labels do not find their way into the wrong hands. One of Scotland's biggest printers of wet-glue labels supplied many of that country's distilleries and reckons more than half of all Scotch Whisky labels are wet-glue. For all types of spirits label, printing and finishing is complex, involving up to twelve operations used to make clothing, carpets and other products. To permit recycling, the label must be easily removable. This is most frequently done by flotation. A leading materials producer now makes recyclable sleeve label film.

Ecologists (rightly) condemn the pollution caused by plastic drinks bottles of all kinds. It is the label industry's responsibility to work together with the plastics industry to improve recycling rates and above all to ensure that the label separates easily and does not pollute the recycled product.

POINTS TO REMEMBER:
- Beverage labels have high growth potential, particularly for self-adhesive technologies
- Bottled water and beer markets are concentrated (except for craft beers), wine market is highly fragmented
- Content of label must conform to local regulations in each country where it is marketed
- Spirits, and spirits labels, need protection against counterfeiting
- Ecological concerns (recycling) are a major factor.

Chapter 4

Labels for cosmetics, health and beauty

Smaller than the food and beverage label market, the cosmetics sector is characterized by high value-added labels. Regulation by public bodies is less stringent, but counterfeiting is common. Cosmetics labels have an important role to play in combatting product piracy.

MARKET SIZE AND MAIN CHARACTERISTICS

In this chapter we look at the whole cosmetics, health and beauty markets, excluding only OTC drugs and any product for which a doctor's prescription is needed. The chief characteristic of this category of products is that unlike prescription pharmaceuticals or basic foodstuffs, they are all products of choice. Accounting for just 9 percent (by volume) of the total self-adhesive market, (and a slightly lower percentage of total label use), they are frequently high value-added decorations designed to sell a high-margin product. Moisture-resistance is a key requirement for many, if not most cosmetics labels. Each sub-group of products (hair coloring, skin cream, shampoo) has its individual labeling imperatives.

Less coercive than food labeling laws, cosmetics regulations and directives are nonetheless binding on the manufacturer, and the label converter needs to know what they are. The main regulations are given below.

DIRECTIVES AND REGULATIONS GOVERNING COSMETICS LABELING

Cosmetics marketed in the United States, whatever their country of origin, must comply with the labeling requirements of the Federal Food, Drug, and Cosmetic (FD&C) Act, the Fair Packaging and Labeling (FP&L) Act, and the regulations published by the Food and Drug Administration (FDA) under the Authority of these two laws. The FD&C Act in particular was passed by Congress to protect consumers from unsafe or deceptively labeled or packaged products.

The main rules laid down by US authorities for cosmetic labels cover basics such as weight/volume (using only US measures) and supplier's name but also determine that a label may not be 'false or misleading'. The label must also warn of 'consequences resulting from the intended use'. Front and back labels must each contain certain information, and instructions for use and all other information must be clearly legible (and in English). For ingredients, regulations lay down in detail the order in which they must be listed, also the type size and prominence. In a curious exception, US laws do not consider soap as a cosmetics product. In another exception, cosmetics do not have to carry an expiry date.

If experts have not substantiated a cosmetic product's safety, the label must display 'prominently and conspicuously' the words 'The safety of this product has not been determined'. Marketing a product with that on the label might be considered a challenge. For aerosols, a whole range of other

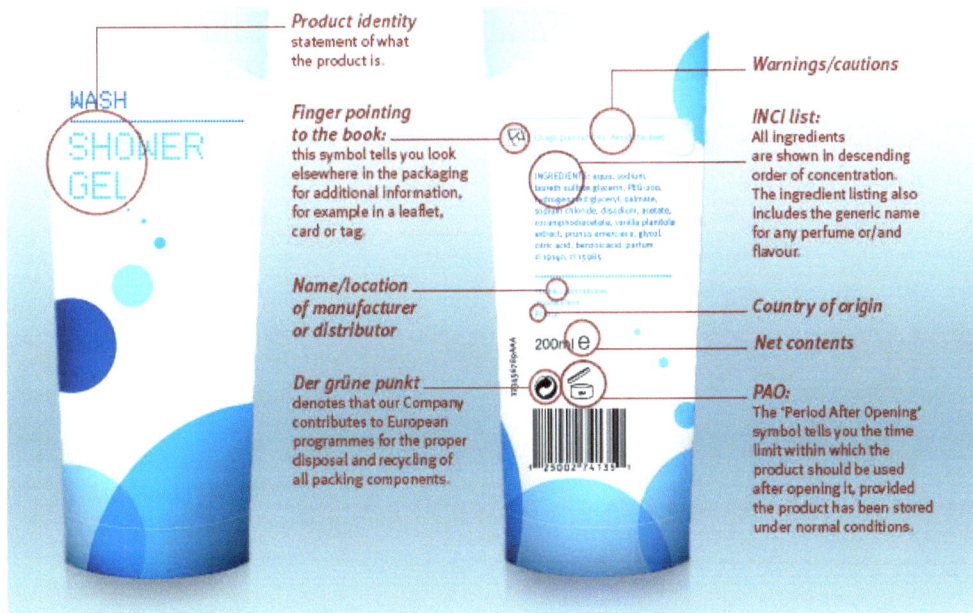

Figure 4.1 What needs, to be shown on cosmetics labels

warnings must figure on the label. Separate and slightly different labeling regulations apply to cosmetics delivered by direct mail. All these US regulations and many others are listed in detail on the website www.fda.gov/Cosmetics/Labeling/Regulations.

In Europe many of the regulations are the same as in USA. Quantities (in metric measures), country of origin and name of manufacturer must all be shown on the label, along with a barcode and, where relevant, a warning (e.g.'Do no throw aerosols into the fire'). Ingredients must be listed using their scientific (usually Latin) names, and the same terms must be used in all countries of the EU (despite the fact that not one consumer in a thousand can understand them). A symbol plus a date shows the 'Period after opening' during which a product can be safely used. Another symbol on the label is used to indicate that additional information is available elsewhere, for example on a leaflet. Ecology-conscious Europeans will often see on cosmetics a symbol indicating that

the label and packaging can be recycled. A recent EU directive (Regulation (EC) No 1223/2009 on cosmetic products) tightens safety requirements and simplifies the procedures for registering new cosmetic products.

To summarize, US, EU and most other countries' laws require the label on cosmetics products to state what it is, what it is used for, its weight/volume, a batch number (for traceability), plus some warning about improper or dangerous use.

Optional information may include (for Europe) a symbol indicating that the product has not been tested on animals. However, a label bearing the words 'We do not test on animals' often means only 'We let other people do the testing for us'.

Similar misinformation on cosmetics labels may include words like 'non-irritating', 'allergy-tested', 'natural' or 'green'. None of these terms has any legal force. The Cosmetics Organic Standard (COSMOS) tries to harmonize organic standards across the globe. To get a COSMOS certification, the product has to meet a strict set of criteria. It ensures that the

L'Oréal	$25 - 30	**Beiersdorf**	$7
Procter & Gamble	$20 - 28	**Shiseido**	$7
Unilever	$15 - 22	**KAO**	$6
Avon	$8	**Johnson & Johnson**	$6
Estee Lauder	$7 - 11	**LVMH**	$4

Figure 4.2 Leading cosmetics groups, annual sales in billion USD. Source: Cab Penhallow Consultancy

product really does contain organic ingredients, and is not just label trickery.

PRODUCTION, DISTRIBUTION AND MARKET STRUCTURES

Statistics on sales by leading cosmetics groups differ widely, no doubt because of ambiguity on what constitutes a cosmetic product.

The listing shown (Figure 4.2 - World's leading cosmetics groups) derives from averaging out statistics of normally reliable sources. Whatever the sale figures used it is clear that in a world market totaling somewhere in the region of 400 billion USD, the three market leaders together account for only some 15 percent market share. This sector is therefore much less concentrated than, for example, the food business.

Procter & Gamble's product range in the cosmetics/health/beauty sector includes leading brands of shampoo (Head+Shoulders, Herbal Essence), diapers (Pampers), shaving products (Gillette, Braun), female hygiene (Tampax, Always) and many popular brands in skin care and deodorants.

L'Oréal is mainly known for its perfumes: its brands include Lancôme, Yves Saint Laurent, and Giorgio Armani, Cacharel, Guy Laroche and Roger & Galet. Its retail division sells the Garnier range of skin and hair care products. It also has specialty brands

sold exclusively to professionals in the hair and body care sector.

Unilever needs no introduction: its cosmetic products include leading skin care brands like Ponds, Hazeline and Caress, shower gels (Lux, Badedas), and eight brands of hair care. It also makes soaps, oral hygiene products and products for baby care.

Distribution of cosmetics and perfumes is via supermarkets for the everyday products like hand cream (Beiersdorf's Nivea cream is said to be the biggest-selling cosmetic product ever), and for many of the brands listed above. Labels for many of these products are ordered by the tens of millions after long negotiations and investigations into the label converter's credentials (financial, technological end environmental).

More exclusive products are more generally retailed in specialty stores, but here again the quantities can be impressive. In many countries, pharmacies are the biggest outlet for slightly up-market brands of cosmetics and hair care. Airport shops are reckoned to account for over 15 percent of certain well-known perfume and deodorant brands as hurried business travellers stock up for themselves or their loved ones with scant attention to the price. A few smaller companies (like L'Occitane) manufacture and sell through their own retail outlets. Avon, which sells only beauty products for women, uses its own unique 'direct selling' network of part-time representatives who sell on a woman-to-woman basis. For the label converter it is useful to note that each distribution channel has its own brands. This is to discourage price comparisons between channels.

LABEL TYPES USED FOR COSMETIC PRODUCTS

Cosmetic products are nearly always branded, and frequently have high value-added. This means that in many cases even very complex, expensive labels are only a very small percentage of the retail cost.

Hair care products must withstand long exposure to water and humidity. If the product is in a squeezable bottle, then the label must not wrinkle. Clear-on-clear filmic self-adhesive labels are commonly used on this category, but aluminum spray cans are more often direct-printed.

```
                    ┌─────────────────────┐
                    │   Main product      │
                    │    categories       │
                    └─────────────────────┘
```

Haircare, shampoo	Hair coloring	Bathroom cosmetics	Skin care, sun protection	Eye makeup, lipstick	Perfumes and deodorants	Shaving products	Oral hygiene

Labels for hair coloring present the particular problem of exact color matching. This is where digital label printing comes into its own, since the same basic product is often sold in fifty or more SKUs each with a slightly different color. For many bathroom cosmetics the label will be as for shampoos. However, look into any bathroom cosmetics cupboard and you will see products in glass bottles, rigid plastic pots, tubes, and rigid or squeezable plastic bottles. The top-selling Nivea (Figure 4.3) is sold in a plastic pot decorated with no less than four labels, of which three are self-adhesive clear-on-clear and the fourth is a circular multi-layer label opening up to give hints for use. Many bathroom products, like Nivea, are greasy, and labels must not discolor, detach or degenerate. Others like nail varnish remover contain chemicals, and these can react with certain filmic label types. The same is true of most skin care and suntan lotions, which can be particularly oily. The peculiar labeling problem for lipsticks and other makeup products is that the labels often need to be very small, but still providing all legal requirements as well as instructions for use and brand logos. Where these products are small tubes (e.g. lipstick) label application can also be a challenge. Perfumes and deodorants cover a vast range from the cheap-and-cheerful up to the eye-wateringly expensive. Direct printing is seen frequently on these products, but self-adhesive labeling is the dominant technology. Printing is often on metallized foils, with spot colors, and finishing generally involves varnish, hot or cold foiling, or embossing. It is estimated that around 70 percent of all perfume labels are on filmic labelstock. For shaving foam or gel the standard decoration is direct printing onto the can or tube.

One leading brand of oral hygiene mouthwash uses front label made to look like a traditional medicine, and a clear-on-clear back label giving ingredients.

Since cosmetics labels are most often displayed without any leaflet or box, the label is the main vehicle to sell the product and to inform on its use. This has encouraged the use of multi-layer or peel-off labels, often used to give complementary information or to include various languages to conform to local laws.

With all cosmetic labels the conditions of storage after purchase are an important consideration for the label converter. As well as water-resistance, many cosmetics labels need to be scuff-proof, and this is not always the case with some digitally printed labels. Since these products must look good on the retailer's shelf, the barcode is generally relegated to a back or contra label, which will also contain all the legally required information (dosage, health warnings, for example). For cosmetics products brand owners will insist that the label and logo must have exact color reproduction. Some brands even have their own special house/brand colors created especially for them by an ink manufacturer. Controlling the supply of these inks will provide an additional means of brand protection.

Every type of label, and every type of print can be found decorating cosmetics products. For reasons of cost, flexo predominates, but there is much use of offset, screen, hot and cold foiling, embossing on luxury brands, metalized or 'soft touch' surfaces and customized finishing. For these reasons the higher-end cosmetics market makes use of combination multi-process presses which, by the nature of their complexity, may also help to deter counterfeiting.

Figure 4.3 Multiple labels are used for this bathroom cosmetic product. Source: Beiersdorf

The flyer in Mexico reads: 'Señora, don't throw them away, we'll buy your empty bottles of perfume. Just call and we will come and pick them up.'

It goes on to offer 30 pesos (about 3 USD) for an empty flask of Lancôme. Hugo Boss bottles for some reason can bring in 3.50 USD.

The offer may be tempting to those with empty perfume bottles lying around, but the sale of the bottles, which are of course destined to be refilled with imitation perfumes and sold as brand-name products for a fraction of the cost, is causing a headache for cosmetics brands.

ANTI-COUNTERFEIT AND BRAND PROTECTION

Product piracy in developing countries may affect even cosmetic products of low value, but serious counterfeiting is confined to luxury items like perfumes – which account for over ten percent of all seizures of counterfeit goods in the US. Perfumes are amongst the favorite luxury goods sold via Internet or Online exchange forums. The ruling handed down by France's Tribunal De Grande Instance in the case of Louis Vuitton v eBay in 2008 is a good example of harsher penalties towards fakes. The court ordered eBay to pay 5.7 million EUR (6.2m USD) in damages after the luxury goods maker, its sister company Christian Dior and four of its brands, accused eBay of illicit sales.

Fake cosmetics are still mostly coming from China, (including Hong Kong, which has some of the worst offenders). However there is hope for an improvement since the declaration two years ago by Jack Ma, boss of China's biggest online retailer Alibaba, that counterfeit goods were 'a cancer' which he vowed to fight.

Fake cosmetics not only harm the brand owners' profits, they can harm the user as well. Counterfeit products have been found to contain cyanide, arsenic, rat poison, urine, and lead, none of which should be finding its way onto, and through, the user's skin. Anti-counterfeit labels and closures must be tailored to the retail price of the product, but modern track-and-trace technology, using QR code/

smartphone technology means that even mass-market cosmetics can be protected. Special print and non-standard colors can make the counterfeited product stand out. Holograms can be faked of course, but they still act as a deterrent. The multi-layer label is also considered to be a useful deterrent since counterfeiters find this more difficult to reproduce. Most leading cosmetic brands have teams of detectives whose sole job is to sniff out fake products sold via Internet, or at street markets.

The closure is often the weakest point of any cosmetics packaging, and though tamper-evident closures are becoming the norm, they will not defeat every counterfeiter (see box).

PARTNERSHIPS TO BEAT THE COUNTERFEITER

The creation of the more luxury cosmetics, health and beauty products labels is often done on a partnership basis, between the brand owner, the designer, substrate supplier, ink manufacturer and converter, so as to make the labels as unique as possible and to deter attempts at counterfeiting. There have even been instances where a press has been uniquely built and configured just to produce one brand owner's dedicated brand product labels.

DIGITAL DEMAND

One major shampoo manufacturer in France has installed, next to its plant, its own packaging and bottle forming plant, making 300 million bottles per year. It also controls its labels which it buys in bulk to reduce costs. But the text on shampoo bottles changes regularly, depending on the laws of the countries where they are sold and in the case of special promotions. This can lead to labels being thrown away. One of the latest projects is therefore to print labels in two phases. This brand owner now asks its suppliers to print and stock the basic labels, and print the text at the last moment, just before delivery. The expected gains are around 600,000 EUR (650,000 USD) per year. Many other label end-users adopt this technique, which favors converters equipped with digital presses for fast delivery of short runs.

ENVIRONMENT, SUSTAINABILITY, ETHICS

Leading cosmetics or perfume brands are keen to use their labels to promote their eco-values. One such company is L'Oréal: after it launched its sustainability platform 'Sharing Beauty with All', a leading US-based labelstock producer offered to do a Greenprint assessment for L'Oréal Americas. Greenprint was introduced in 2010 and is an environmental lifecycle assessment tool. It measures the environmental impact of labels across the six categories: fossil material, trees, water consumption, energy, greenhouse gases and solid waste.

As a result of the assessment, L'Oréal Americas is transitioning the labels for some of its leading

> **Not really a label (though it could be used as one) is a UV sensor which changes color according to the strength of the sunlight. Linked to an app, it recommends what strength of suntan lotion should be used depending on local weather conditions.**

products to one of Avery Dennison's extra-thin products. By providing twice the number of labels per roll, this labelstock produces 40 percent less waste and also reduces the converter's energy bill. As an added bonus, the 33-micron PET film used by L'Oréal is fully recyclable. This initiative by L'Oréal (now followed by many other brands) also extends to the use of eco-friendly inks on the labels of its products.

POINTS TO REMEMBER FOR COSMETIC LABELS

- Huge, expanding market sector for luxury labels
- Subject to strict regulations in most countries
- No-label look is increasingly popular
- Color, appearance, water and oil resistance are key label considerations
- Multiplying SKUs and shorter print runs encourage digital label printing
- Multi-layer labels are much used giving added information or various languages
- Product piracy is a problem: the label can help to solve it
- Ecological concerns are important.

Chapter 5

———

Pharmaceutical and medical sectors

———

Worldwide revenues from pharmaceuticals are estimated at around one thousand billion USD. The United States market represents just under half of this total. Globally, pharmaceutical markets are growing fast and have almost trebled since 2000. Growth in certain emerging markets (like China) is especially fast.

———

MARKET CHARACTERISTICS

In both developed and emerging countries, health care budgets are expanding. Awareness of hygiene and health factors encourages the label and packaging sector to invent better solutions to health-related issues. While no reliable figures exist for the global pharmaceutical label and packaging markets, they are of the order of 50-60 billion USD - and are forecast to reach more than 80 billion by 2020.

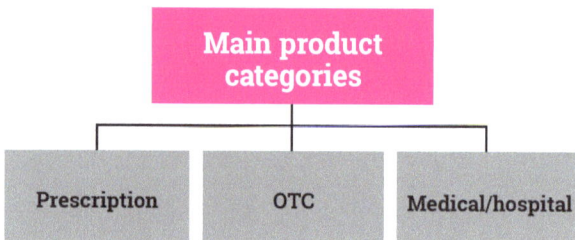

Main product categories

Prescription	OTC	Medical/hospital

Pharmaceutical and medical products can be broadly divided into prescription drugs, non-prescription (or over-the-counter/OTC) drugs, and hospital equipment/consumables. While these categories are partly overlapping, their requirements for the label converter are substantially different, for legal, usage and marketing reasons. The principal

regulations, which the label designer or converter needs to know about, are detailed below.

DIRECTIVES AND REGULATIONS

The two main regulatory bodies are the Food & Drug Administration (FDA), for the United States, and the European Union for Europe. Other official agencies regulate drug sales in China and other countries but generally, their rules tend to approximate to those of the FDA and EU.

Label converters are not expected to know all the very complex regulations regarding the text for pharmaceutical labels but should at least be familiar with the general requirements so as to be able to discuss with their customers and understand their constraints. For the United States, the Federal Food, Drug and Cosmetic Act (FFDCA) defines 'labeling' as all labels and other written, printed, or graphic matter upon any article or any of its containers or wrappers, or accompanying such article. The term 'accompanying' is interpreted liberally to mean more than physical association with the product. Label converters wishing to get full details of US compliance requirements can consult the US Registrar Corp who (for a fee) will provide confirmation that a given label or package meets FDA requirements.

Company	Global sales 2014 (USD Billion)	Company	Global sales 2014 (USD Billion)
Novartis	49	Johnson & Johnson	32
Pfizer	46	Glaxo Smith Kl	30
Roche	39	Astra Zeneca	26
Sanofi	36	Gilead	24
Merck	36	AbbVie	20

Figure 5.1 Top pharma companies by global sales. Source: PM Live

PRODUCT INFORMATION REQUIRED ON PHARMACEUTICAL LABELS/PACKAGING

Requirements will vary from country to country and product to product, but the following will generally be required:

- Product name
- Ingredients, active and inactive
- Drug information
- Purpose and use
- Any specific product warnings
- Directions for taking or use
- Information about allergic or other reactions.

MARKET STRUCTURE – PRODUCTION, DISTRIBUTION AND MARKETING

The pharma business has seen increasing concentration in recent years as bigger firms bought up smaller ones to increase their market share. Globally, the biggest drug producers are those listed in Figure 5.1.

You will notice that despite the huge size of leading companies, the market is still not concentrated. Despite the strict regulatory framework existing in most countries, smaller pharmaceutical companies exist in their tens of thousands

IMPORTANCE OF CONTRACT PACKERS IN PHARMA SECTOR

Many pharmaceutical companies do not pack their own products. The packing/labeling (and sometimes even the preparation of the product) is given over to specialized contract packers. This means that the label converter has to conform to the demands of both the drug company and the packer. Contract packers are frequently called upon to pack small quantities for clinical trials. This is why many of them have invested in digital label presses. Some, like Netherlands-based Sinensis, specialize in providing short-run packing services for pharmaceutical manufacturers.

US-based Catalent has 25 packing sites covering North America, Europe, China and Japan. However, although the pharmaceutical business is global, contract packers are mostly not. GMP Pharmaceuticals for example covers mainly the Australian and New Zealand markets, Redditch Medical does business mostly in UK.

MARKETS AND PRODUCT TYPES

With few exceptions, the doctor, and not the patient chooses a prescription drug. It is therefore one of the few products for which the label/packaging has almost no marketing function. It is almost unique in that in nearly all developed countries, the 'buyer' either gets it free, or pays only a nominal amount. Labels for prescription drugs therefore tend to be one or two colors only. In some countries a two-layer label enables users to peel the top layer and stick it to their reimbursement form, however with increased 'paper-free' transmissions this practice is on the way out. Either the label or the packaging will need to carry a braille description (this is also the case for many OTC medicines). If braille is carried on the label it is generally done by embossing, although screen printing and digital printing are also used.

Users of non-prescription (or OTC) drugs are free to choose the products they buy, so normal competitive forces come into play. Frequently branded and supported by heavy advertising, they need to

Figure 5.3 Syringe label from Schreiner Medipharm

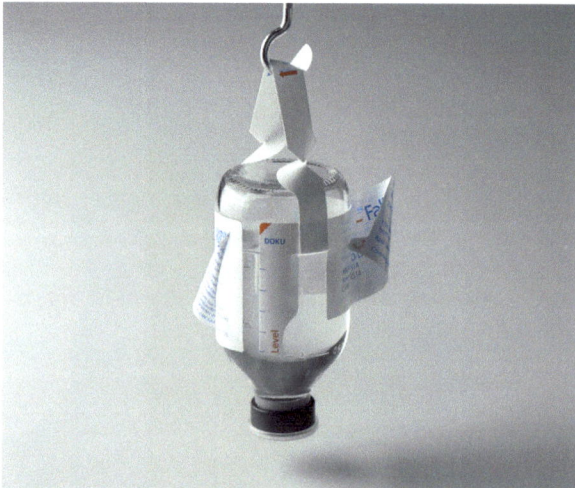

Figure 5.2 Not just a label – August Faller has developed imaginative label solutions for the medical sector

stand out on the pharmacy shelves or, if not displayed, at least be attractive in appearance when bought, or when visible in a bathroom cabinet.

The medical and hospital sector is where many label innovations find their use (see Figure 5.2). Consumables include blood pouches, serums and indeed all products administered to, or injected into, a patient. The double or triple layer self-adhesive label, with one copy going onto the patient's record, avoids the risk of errors when copying by hand. For blood bags, the hospital sector needs very special labeling materials due to regulatory requirements and the extreme conditions associated with blood storage and processing. Several labelstock producers have developed laminates specifically for this use. UPM Raflatac for example have designed a range of primary blood bag label materials fully compliant with ISO 3826 standards. These labeling products use an adhesive that can tolerate temperatures from +121 deg C during steam sterilization, down to -80 deg C, as well as freezing in ethanol to -50 deg C.

Other medical equipment requiring specialty labels includes syringes and all kinds of flacon. Much ingenuity has been shown by label converters whose innovations have included 'Needletrap' a safety label

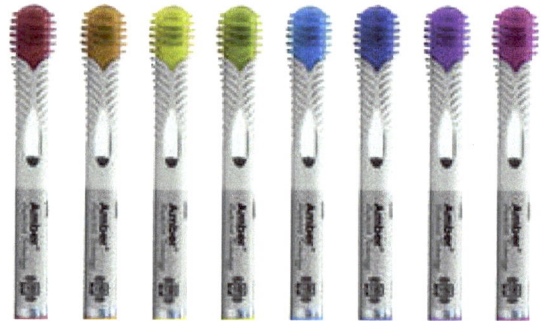

for syringes and 'Flexicap' a protective pack with integrated label for the protection of dangerous fluids. A particularly ingenious syringe label has been designed for so-called 'auto-injector' syringes, as used for example by those suffering from acute diabetes (Figure 5.3). As well as the usual printed information, this label comes with an anti-slip varnish to prevent slipping during injection, an integrated hologram which identifies the product as an original, and a printed QR code or Near Field Communication (NFC) chip which provides a link, via smartphone, to a website with additional product information.

Hygiene, inventory accuracy, patient safety: medical facilities and hospitals have to follow many regulations regarding clinical practice and keep track of their medical equipment at the same time. For this, RFID labels are being used, particularly in hospitals and clinics. In a Danish dental clinic, RFID tracking of dental instruments is said to improve patient safety and hygiene. Instruments can be tracked continuously from the time of entry into the system, through their use and processing, and can even be sterilized (without harming the label).

For label converters, the importance of the hospital/medical sector must not be underrated. Hospitals and health services are quite rightly more concerned about reliability than cost. Having established a relationship of trust with a label supplier, they are reluctant to change. That is the reason why so many world-class label converters (CCL, Reynders,

Schreiner Medipharm and others) all have a division devoted exclusively to meeting the needs of this sector.

LABEL TYPES

Pharma and medical labels can be of many types, but self-adhesive and to a lesser extent sleeve labels dominate. Frequently all component parts of a label (paper/film, backing, adhesive, inks, coatings) must conform to strict specifications. Peelable (multi-layer) self-adhesives, as mentioned above, are increasingly used for hospital applications. In many countries, blister foil packs are the primary packaging for pills, and these are generally printed on the same flexo or digital presses as for labels.

PRINT TECHNOLOGIES, INKS AND ADHESIVES

All print technologies can be used for pharma/medical labeling. Increasingly, digital printing is being used because of the need for short-run test batches of drugs, or for individualized labels for hospital use. However for nearly all label applications in this field the biggest hazard is migration. Harmful substances which might be tolerated in inks or adhesives used for other purposes cannot be tolerated where there could be a risk to a patient's health.

WHAT IS SPECIAL ABOUT PHARMACEUTICAL AND MEDICAL LABELING?

Special features can be grouped under four main headings:

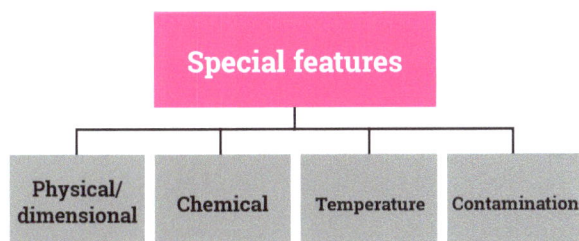

Special features
- Physical/dimensional
- Chemical
- Temperature
- Contamination

BRAILLE

Different countries use different types of braille so the printer needs to know the rules of the country where the braille will be read. Braille for pharmaceutical artwork generally uses 'uncontracted' braille (where each letter or character of the braille alphabet is represented by a corresponding braille character). So for the most part it is a simple matter of using an appropriate braille font. However, there are certain rules to be followed regarding braille font indicator characters, such as the number symbol, letter symbol and capital letter symbols. The EBU European Braille Code has been accepted by some European countries for use on pharmaceutical packaging, and for the UK market, the Royal National Institute for the Blind (RNIB) will provide specifications for dot height and style. Further details on braille fonts can be found on the Pharmabraille website (www.pharmabraille.com). Label converters have the option of embossing the braille lettering (reckoned to be the most reliable method) or using screen printing. Some digital presses like the recently developed Atlantic Zeiser Braillejet™ print digitally using a high velocity UV ink to achieve immediate fixation of the dots. Whatever the method used, care must be taken to respect regulations on the height (and hence the readability) of the braille dots. Once samples to correct specifications, layout and materials have been produced it is usually necessary to seek approval from the pharmaceutical governing body relevant to the country in which the products will be sold.

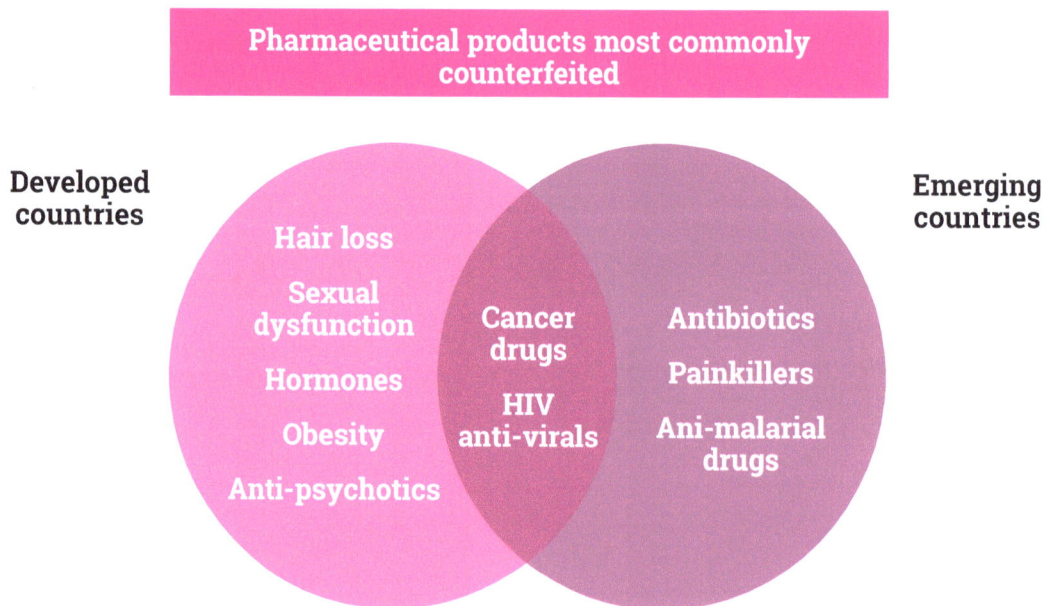

Pharmaceutical products most commonly counterfeited

Developed countries

Hair loss

Sexual dysfunction

Hormones

Obesity

Anti-psychotics

Cancer drugs

HIV anti-virals

Emerging countries

Antibiotics

Painkillers

Ani-malarial drugs

Figure 5.4 Pharmaceutical products most commonly counterfeited

Physical:
Labels must be able to withstand rough handling, so they must be scuff-proof, but also water- and humidity-proof. This means that the inks must be fully water-resistant, even when patients drop their medicines into the bath. This applies equally to the labelstock/adhesive used, which must resist humidity and ageing. The typeface (font) needs to be legible for the intended user (who might be an elderly person with impaired eyesight). In Europe and many other regions braille lettering is compulsory (see box). Both EU and US regulations stipulate font size and legibility. For the EU in addition a directive (Article 63(1) of Directive 2001/83/EC) required that 'labelling and package leaflet shall appear in the official language or languages of the Member State where the product is placed on the market'. This requirement has greatly increased the use of multi-page or booklet labels with one page for each language.

Chemical:
In both pharmacy and hospital usage, labels may come into contact with hazardous or corrosive liquids, and must remain legible under even extreme exposure. The print must be indelible, and must not fade with age or exposure to sunlight.

Temperature:
The same problems arise as in the food sector; only here the conditions can be even more extreme. Sterilization, frequently by autoclave, is a well-known difficulty, requiring the label converter to use only heat- and moisture-resistant adhesives and inks for an environment with temperatures over 100 deg C. In some cases sterilization may also use extremely cold temperatures, and storage – of vaccines for example – may mean the label has to resist weeks or even months of cool or cold temperatures which may go down to minus 80 deg C where the physical properties of adhesive, film and paper can alter. This

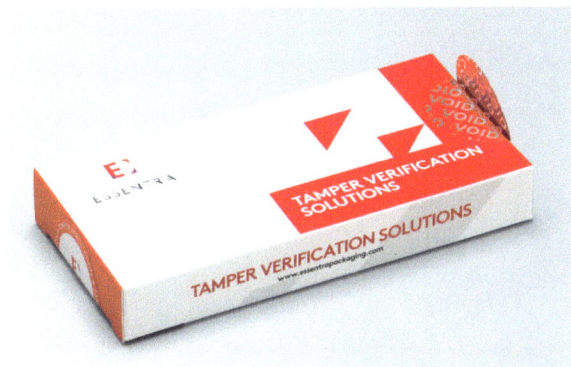

Figure 5.5 Essentra is one of many suppliers providing a range of tamper-evident solutions

is a major challenge for the label converter. However, as described above, major labelstock manufacturers have developed materials suitable for these extremes, and inks are also available for such specific applications.

Contamination:

This is the ultimate danger with all labeling and packaging in the medical sector. Harmful substances can permeate through the label and into the contents which is why here, as in the food sector, migration of chemical substances in the ink or adhesive of the label can cause problems. However in the case of most solid pharmaceuticals (e.g. pills) the substance itself is encased in a blister pack which is itself rarely labeled. More important among potential contaminants is sunlight, which can pass through the label and adversely affect the product. However other forms of contamination can occur when labels are applied, for example, to blood bags.

THE BATTLE AGAINST COUNTERFEITERS

The above risks concern problems which may happen accidentally in pharmaceutical labeling. More frequent, unfortunately, are the problems of fighting against deliberate counterfeiting (worldwide, Viagra is said to be the most frequently counterfeited medical product – see Figure 5.4). Other commonly counterfeited products include antibiotics, cancer drugs, anti-

malaria pills and treatments said to cure conditions ranging from hair loss to obesity or depression. In some developing countries medical authorities reckon that more than half the pharmaceuticals in circulation are counterfeit. As a result millions of people are at risk of buying drugs that in the best case will not cure them and may even do deadly harm. Labels and packaging have a vital role to play in fighting this menace. Methods (see figure 5.6) include:

- Overt (visible) brand protection involves visible/tactile security features that can be easily read by the public and by pharmacy employees. This involves holograms, special colors, hard-to-copy digital watermarking, 3D features, encrypted RFID, smart-phone-readable QR codes and above all reliable, tamper-evident closures (Figure 5.5)
- Covert (invisible or masked) brand protection using invisible inks or other security features needing UV light, special penlights or other special equipment to detect. Covert protection also includes the use of special papers or cartonboard containing taggants
- Forensic security features: this category includes highly complex security features needing scientific evaluation. Since such tests mostly need time to be completed, such forensic protection is generally reserved for high-value items with a high risk of counterfeiting.

All these security features have a cost, which needs to be proportional to the risk, and to the value of the product (Figure 5.6).

IMPACT OF NEW EUROPEAN DIRECTIVE ON SERIALIZATION: 2011/62/EU ('FALSIFIED MEDICINES DIRECTIVE')

Serialization: what is it? Serialization means giving a unique, traceable number to each individual pack of a product, along with corresponding numbers to each carton or pallet. As an anti-counterfeit system it is not limited to pharmaceutical products, but will be used worldwide for these products which are particularly subject to counterfeiting (see Figure 5.5).

For Europeans, compliance with the 'Falsified Medicines Directive' (2011/62/EU) will be mandatory from 2018 and will cover all sales of pharmaceutical

Diagram shows the relative costs of authetication devices without taking into account the costs of hardware validation infrastructure (readers)

Higher

Price

Encrypted RFID

RFID

Reflection holograms

Holograms

Refraction holograms

Security papers and substrates

Optically variable

Security links

UV and Fluroescent features

2D barcodes & dataglyphs

Tags and taggants

Surface feature authentification

Mass serialisation

Digital watermarking and encoding

Low

Security level

Higher

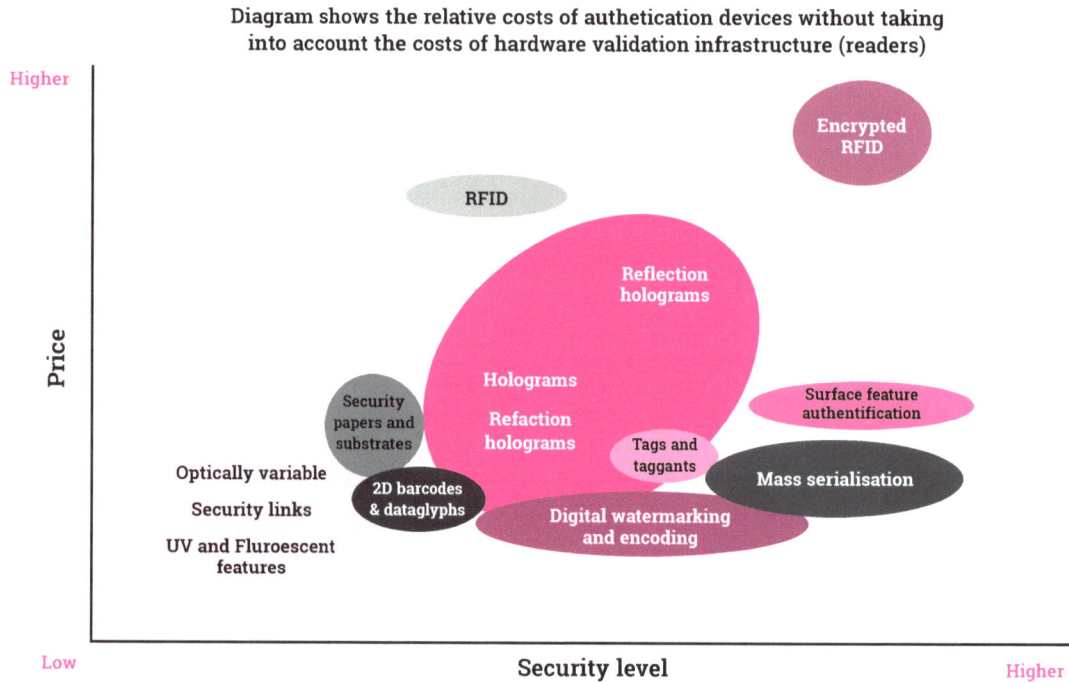

Figure 5.6 Relative costs of authentication devices

products in the European Union, whatever their country of origin. Large markets like China have already implemented a law that requires serialization of each single product including full traceability over the entire logistics chain. Other countries like the USA will be among many others following very soon or at least starting to think about this.

While current serialization requirements are limited to marking the unit of sale with a unique data carrier, by 2023 the process will require a product to be traceable through the entirety of its journey - from the carton/pallet through to the retail pharmacist or hospital. In the United States, the Healthcare Distribution Management Association (HDMA) is suggesting that pharmaceutical companies should begin to support this level of serialization, called aggregation, now. The problem facing the pharma label converter is that no one knows how the different markets of the world will specify the

rules for serialization. A leading company in this sector is Atlantic Zeiser, whose CEO went on record as saying that we should 'Focus on the installation of a central serialization software and database application that is characterized by a modular architecture in order to be able to grow seamlessly with any need. This way it is possible to quickly adapt workflows and coding principles to new emerging regulations without the need of reprogramming and time-consuming revalidation'.

Another international company developing software for unique product identification (UPID) and track & trace solutions is Adents. This company, based near Paris, has worked with Microsoft to jointly develop and market a new Cloud platform. Named Adents Prodigi, it has been hailed as the only Level 4 traceability solution that can centrally manage regulatory requirements imposed on the pharmaceutical industry.

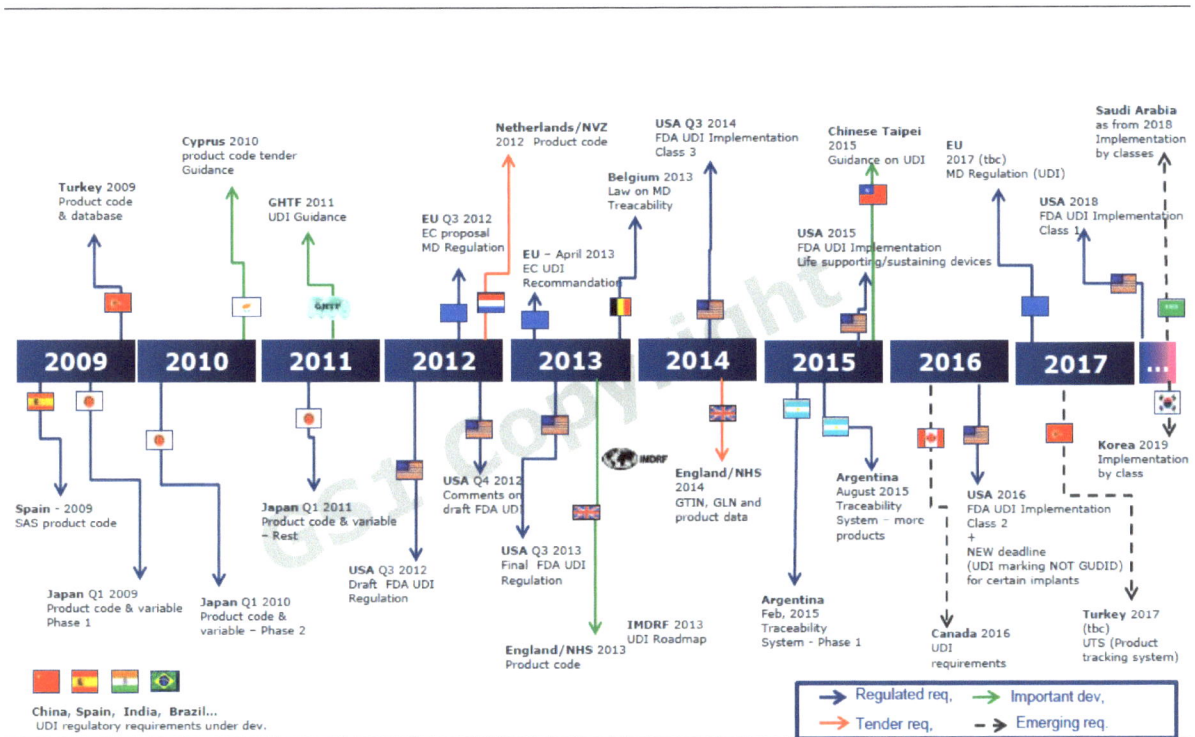

Figure 5.7 The road to serialization is not straightforward! (source: GS1)

GRAY MARKET DISTRIBUTION

A particular problem of pharmaceuticals concerns the selling of genuine goods by unauthorized persons or companies. For example, an importer in a country with artificially low prices for certain medicines may try to re-sell them in a high-price country where they will compete unfairly with the same goods sold by an authorized distributor. The brand owner may need covert protection, which can be given by special inks which change color in UV light. In many cases brand owners may just need to track cross-border movements of their products to trace trading movements which while unwanted may not be illegal.

WHO MAKES WHAT?

The constraints and regulations governing pharmaceutical labels are so strict that many

converters specialize only in this end-user sector. Specialized label converters with interesting new ideas for pharmaceutical labeling include:

CCL – probably the world's largest self-adhesive label supplier, CCL has 23 plants worldwide exclusively focused on the pharmaceutical industry. All of these plants offer serialization, plus a high level of security control.

Constantia Flexibles – also a major player. Constantia's label division makes not just labels but the full range of packaging requirements for pharmaceuticals.

PagoPharm – based in Switzerland, makes multi-layer, security and functional labels.

Reynders Pharmaceutical Labels – this division of the leading Belgian label group is particularly specialized in anti-counterfeit features, and in fully

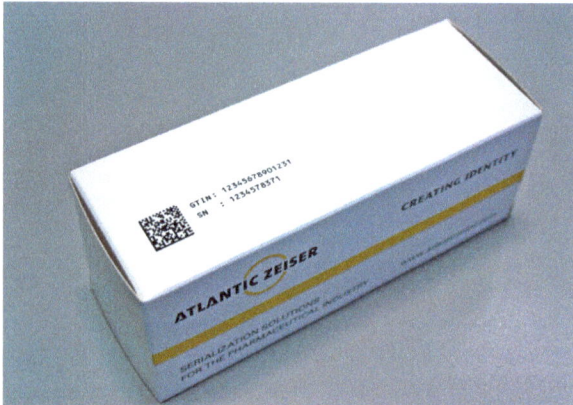

Figure 5.8 Atlantic Zeiser's Medtracker is a tool to manage serialization

integrated camera inspection systems.

Schreiner Medipharm – with plants in Germany and USA, has developed a range of innovative products to simplify the dispensing and administration of pharmaceuticals, particular for hospital and laboratory use. Examples are the 'Needle trap' (photo) and the protective closure and packaging for dangerous liquids (photo).

Securikett – this Austrian label converter, as its name implied, is specialized in security labeling and track-and-trace systems, mainly for the pharmaceutical and medical sectors. Stratus – this Franco-Belgian converter makes labels for all end-user sectors, but has six production lines reserved exclusively for pharmaceutical labels.

Autajon and August Faller – market-leading label and packaging producers based in France and Germany respectively also have plants producing folding cartons, leaflets and labels for many different products, but both strong is the pharmaceutical/medical sector.

POINTS TO REMEMBER
- Strict quality compliance rules apply for pharma labels
- Although leading drug companies are big, the global market is not concentrated
- Labeling requirements are different for prescription drugs, OTCs and hospital/medical requirements
- Many pharma labels must stand up to scuffing, to chemicals and to extremes of temperature
- Serialization will, it is hoped, reduce the very widespread counterfeiting of pharmaceuticals.

Chapter 6

Non-food retail products and household durables

Most products in this category have low unit value and do not need complex anti-counterfeit measures. Among the exceptions are high-end luxury household goods, designer clothing, and shoes. For bathroom or kitchen products a damp-resistant label is essential. This chapter also covers clothing labels. Smart technologies (RFID, NFC) are a common feature in identifying and tracking clothing of all kinds.

MARKET SIZE AND CHARACTERISTICS

Reliable estimates show that non-food retail and household durables account for one in five of all self-adhesive labels. Statistics are less available for other types of label (notably wraparounds), but it is probable that the percentage is similar. Taking all label types together, the global market for this category of labels is of the order of seven billion USD.

Non-food retail products include 'under-the-sink' products for cleaning and disinfecting. Some of these are hazardous, and therefore require special warning labels. We also include all kinds of apparel and shoes. These account for the biggest volumes of labels, mostly fabric labels. Also included in this category are most of the goods you will find in a hardware or do-it-yourself store, or in a sports goods retail outlet.

Household durables like washing/drying units may be labeled by the manufacturer and by the retailer, and (for some regions) must carry labels showing their energy efficiency. For some types of household durables, smart labels can provide extra functionalities and value added

For all the various products in this category, a very wide range of labeling materials is used. For many of

them, resistance to damp is important, so paper labels and certain adhesives may not be suitable; for clothes labels the requirements include a print method that can withstand repeated washing (often in very hot water) or dry-cleaning.

Other products (e.g. plants, garden equipment) may need highly durable labels able to resist exposure to rain or sun without fading or peeling off.

For all kinds of clothing, woven and care labels represent a huge market, as do swing tags. RFID labels are now becoming the standard way of controlling supply chains.

Most products in this category have low unit value and do not need complex anti-counterfeit measures. Among the exceptions are high-end luxury household goods, designer clothing, and shoes.

With ever-expanding variety of Stock-Keeping Units (SKUs) and shorter delivery times, digital printing is increasingly used, especially for flower and plant labels, but also for other label types where pre-launch runs need small numbers of labels.

Regulations and standards vary widely according to the type of product and the country/region in which it is to be sold. The following paragraphs give some of the more important directives and regulations affecting labels.

REGULATORY BACKGROUND: ENSURING PUBLIC HEALTH AND SAFETY

Labeling regulations for non-food retail products vary greatly according to the type of product, but must in all cases give the quantity (in metric for Europe, in locally used measures for US and other markets), the composition or ingredients, and what it is to be used for. If the product is, or could be, dangerous, this must be stated as well. For many products in this category, pictogram labels are often used to highlight dangers or denote the materials used in making the product.

The European Union in particular aims to use the label to safeguard the health, safety and interests of consumers. This policy promotes consumers' rights to information and education, and their right to organize in order to defend their interests. These citizens' rights have on several occasions been used to challenge labels and ads held to be inaccurate or illegal.

Different labeling rules apply to each category of hazardous or environmentally challenging product. Detergents for example can contain ingredients - surfactants - that make them clean more efficiently but may damage water quality when released into the natural environment. As such, their use is carefully controlled, and labels on detergents must give details of recommended dosages for different washes in a standard washing machine.

The REACH legislation in Europe is a system for the registration, evaluation, authorization and restriction of chemicals. Specific labeling requirements apply to all such products which may harm the user, the general public or the environment. The label must specify the following information:

- Name/address of supplier
- Name of the substance or mixture and/or identification number
- Nominal quantity of the products
- Hazard pictograms (graphic composition combining a symbol and another visual element)
- Signal words ('Warning' or 'Danger')
- Risk phrases (for example 'Fire or projection hazard', 'Fatal if swallowed')

- Safety advice (for example 'Keep only in original container', 'Protect from moisture', 'Keep out of reach of children').

In some cases, the law may require tamper-evident closures.

Note: the term 'Ecolabel' is used more and more. Ecolabels are not strictly speaking labels, but are certifications awarded in Europe to products and services which have a lower environmental impact than other products in the same group.

In the United States the Toxic Substances Control Act—TSCA and the Federal Hazardous Substances Act (FHSA) are the main regulatory instruments governing the composition and labeling of dangerous products. They are both administered by the Environmental Protection Agency (EPA). Much overlap exists among these various US laws. Labeling requirements are if anything more exacting than their equivalents in Europe. US hazardous labels must for example give an ingredient statement, use classification, precautionary labeling (warnings to consumers), work protection labeling, directions for use, labeling claims, storage and disposal instructions, identification numbers, company name and address data, graphic symbols on the product label, and content and net weight statements.

HOUSEHOLD CLEANING PRODUCTS

For many years domestic cleaning products had minimalist labels since once bought they were kept out of sight under a sink or in a bathroom cabinet. A trip to any supermarket shows this is no longer true. Most retailers sell their own brands of household products in competition with major brand owners like Unilever or Reckitt Benckiser, and eye-catching labels have become essential. Main label types are the following:

Wraparound

The wrap-around label is used frequently for low-margin, high-volume products like demineralized water. It is also to be found in supermarkets where the retailer offers a 'budget' line of household products. It may be paper (cheaper) or synthetic (better water-resistance).

```
                    ┌─────────────────────┐
                    │   Decoration for    │
                    │  household products │
                    └─────────────────────┘
```

Wraparound label	Wet glue label	Self adhesive (paper or filmic) label	Sleeve label	In-mold label	Direct printing (no label)

Wet glue

Still a widely used labeling technology for household products, the wet glue label with its susceptibility to moisture and mold, is losing out to other methods. However, in common with the wraparound label it has the advantage of being inexpensive.

Self-adhesives

Self-adhesive labels account for over half of all household product labels. Improved moisture-resistant inks and varnishes have led to the use of paper as a face material in many cases, but the majority of self-adhesives have synthetic face materials which look better especially after prolonged use/storage in damp conditions. The other reason for preferring synthetics is that most household products come in squeezable plastic containers: if the label is more rigid than the bottle, it quickly becomes wrinkled and unattractive. Traditionally, these squeezable labels were made from polyethylene; however, they can now be manufactured using a BOPP (biaxially oriented polypropylene) film which is conformable, squeezable and printable. A gloss varnish can even be applied to guard against smudging and water damage. Innovia and Jindal are among the film producers specializing in these 'squeezable' films. All the leading labelstock producers, including Avery Dennison and UPM Raflatac, now also offer 'squeezable' face materials.

Sleeves

Sleeve labels, little used for these products five years ago, are now the preferred decoration for one in six of all household products. The reason is that many products are packed in irregularly shaped containers: some have a handle, others a thin neck or a nozzle. A sleeve label covering the whole surface in 360 degrees gives maximum surface for graphics, ensuring that the product will always look attractive on the retailer's shelf. The main types of film used for sleeve labels are mono-oriented PVC, PET, OPS, OPP and TPE films, and for household products the selection criteria are a high degree of shrinkability (to fit around very irregular containers), and resistance to shock and humidity. With a thickness which can be as low as 3 microns, the shrink sleeve is economical on material. However, the printing (usually reverse gravure or digital), forming and application of sleeve labels means they are an expensive option. Shrink sleeve labels often come with a tamper-evident closure, however this does not seem to be widely used for household products. Market leaders in the shrink label sector include Sleever International and CCL, both fully integrated companies providing not just the film but also the seaming equipment and the shrink tunnel.

In-Mold Labels (IML)

IML accounts for only two percent of the total volume of label printing worldwide, but the average growth rate is between three to five percent, with Europe accounting for nearly two thirds of total IML usage.

With the in-mold labeling process, a pre-printed (usually polypropylene) label is placed inside the mold before a plastic container is blown, injected or thermo-formed to produce a plastic bottle or tub. This process has been in use for many years, mainly for decorating high-volume dairy products. However today it is estimated that 20 percent of IMLs are used for under-the-sink household products. The advantage for the brand owner is the 360 degrees of coverage that IML provides for promoting the product, and the security of

Examples of successful IML

A Belgian household products company recently switched its bleaching powder product from carton packaging to plastic containers decorated using IML. The change to IML is reported to have increased sales by adding to the customer value of the product. In North America, where IML is not so well established, the Canadian company IML Labels supplies mostly to the food sector but notes rising sales for household products, because of 'the shelf impact, with the full 360-degree decoration and eye catching graphic quality.'

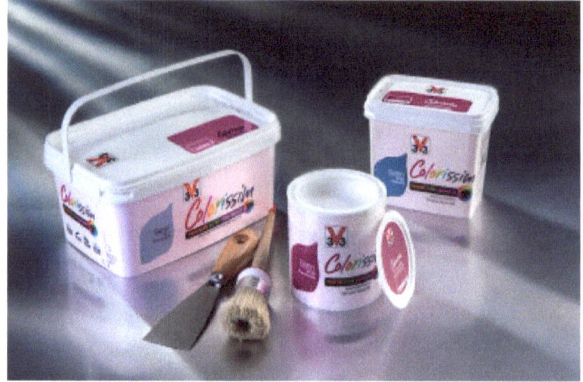

Figure 6.1 IML provides an attractive label solution for paint

knowing that the label cannot become detached from the product. For the final consumer the advantage is a re-closable and (even) re-fillable container with no risk of seepage either into or out of the contents. This is particularly important in countries where corrosive and toxic products like bleach are sold in plastic bottles which when empty can be refilled from sachets of concentrated product, such that the same plastic bottle can be refilled many times.

Direct print

Direct printing though perfectly feasible for plastic containers, is not much used for household products. However as recently as drupa 2016 Agfa unveiled new technologies for direct printing, aimed particularly at the food and home appliance sectors. Krones also, though mainly in the beverage filling/labeling business, is also developing digitally printed direct print technology. Direct printing is a potential threat to the growth of the label business.

PAINTS, VARNISHES AND WOOD STAINS

The particular challenge for these products is to ensure that the color swatch displayed on the label matches the contents of the pack. Label converters have traditionally achieved this by delivering one standardized label for each size of container, plus a roll of single color swatches for each color in the customer's range of paints or stains. This leads to wastage, since the volume of paint sold in each color is unknown. Also the result is not aesthetic and fails to stand out on the shelf. Several manufacturers have gone over to using digitally printed labels, which give a more professional finish to the product, and also reduce wastage as the labels can be ordered and the pails made up and delivered at very short notice. One French paint producer uses in-mold-labels (IML) protected by a UV varnish, on pails made of injection-molded polypropylene. The result is 'appealing and attractive' for the home decorator, but too costly a solution for professionals and artisans, as each color of paint and each size of pail has its individual IML label. For all products of this type, whatever the product decoration method adopted, color coordination is paramount. Color Management Systems (software system used to ensure color consistency among different input and output devices so that printed results match originals) have gone a long way towards solving color coordination. Companies such as X-rite (part of the same group as Pantone) and GSE are just two of the many specialist suppliers offering color management systems to the label industry.

SPORTS GOODS AND OTHER LEISURE PRODUCTS

Even mass-market sports goods are frequently branded, and require labels which conform to legislation/language regulations in the country of destination, offer protection against product piracy, and yet are cheap. Where a textured finish is needed, heat transfer labels are being used for sports goods (as also for consumer durables). They offer excellent durability. Sports shoes, a mass product, suffer severely from product piracy.

While labeling mass-production sports items needs to be cost-conscious, equipment for professional sportsmen (or for those who emulate them) must meet exacting standards, and must be protected from counterfeiters. Wilson, a leading maker of ball sports equipment, is using an RFID-based labeling system to identify each carton shipped out of its factories. Using printers/encoders supplied by Zebra, Wilson aims to achieve anti-counterfeit protection as well as track and trace capabilities for total asset visibility throughout its supply chain. When the new SAP system is complete, Wilson plans to integrate the RFID label data further into its operations so that it can encode additional information, such as ship-to address, SKU number and item quantity.

LABELS FOR LUXURY GOODS

A high unit price (e.g. for handbags or luxury items of clothing) attracts counterfeiters like flies around a jam pot. Labels have a major role to play in helping importers, retailers and consumers to establish authenticity (Figures 6.2 and 6.3). To replace the QR code, secure identification specialists offer an item level unique ID label enabling anyone along the supply chain (including customs officials) to authenticate the product by means of a smartphone.

It will come as a surprise to most people that, measured by the number of customs seizures, shoes are the most smuggled product of all. A report by the World Customs Organization puts Nike in the unenviable top slot for the biggest number of counterfeit goods customs seizures, well ahead of Apple (second) and Rolex (third). Adidas occupied the fifth slot. Manufacturers of luxury footwear have resorted to electronic devices to bring this

Figure 6.2 Beating the counterfeiter: customers' authentication

Figure 6.3 Visible counterfeit deterrents

counterfeiting under control (see box)

For shoes of all qualities, as for many products in this category, pictograms are often used to denote the materials used in making the product. For footwear, pictograms on the label or tag are used to denote which parts of the shoe are in leather, and which in rubber or other material (for details on European requirements, see Directive 94/11/EC of the European Parliament). A voluntary EU eco-label also exists for footwear. This label helps consumers identify footwear whose life-cycle (production, use and disposal) has a low environmental impact.

Italian shoemaker and fashion house Salvatore Ferragamo conceals a smart label in every left shoe it makes. The RFID tag embedded in the label can be interrogated by a special transmitter-receiver which sends back the signal that the shoe is genuine – or not. The company says most fakes come from China. Ferragamo and Chinese customs authorities seized more than 34,000 counterfeit products in 2015. Ferragamo estimates the value of these goods to be more than 17 million USD. Other makers of luxury goods are said to be using the same or similar devices to trap product pirates.

Figure 6.4 M&S shops use a hand-held reader for inventory control

APPAREL LABELING (1) – WOVEN AND CARE LABELS

In volume terms the woven and/or care apparel label is among the biggest, globally. Almost every item of apparel contains at least one, usually several. Care labels are generally thermal printed in black, and are sewn or affixed in such a way as to be concealed. Woven labels are designed to be seen, and to enhance the value of the item of clothing, so are often colored to show a brand or logo, and made of cotton or satin. They can also be printed, and for this, digital printing is often used for short runs. They are also used for late-stage differentiation of colors (the label is dyed along with the article of clothing). Since all these labels may come into contact with the skin, they must not contain any irritants. Suppliers such as Avery, in cooperation with major retailers, have for example planned to phase out completely adhesives containing the irritant alkylphenol ethoxylate (APE). Both woven and care labels have to be designed to last as long as the item of clothing itself, and to withstand repeated washing and dry-cleaning. A recent addition to this category is the printed transparent thermoplastic polyurethane label, which has a soft feel but withstands high temperatures and repeated washing.

APPAREL LABELING (2) –THE SWING TAG

Swing tags have two functions: either as logistics labels thanks to their QR or barcode, or as a selling aid for the retailer. Sometimes the swing tag function is transferred to a self-adhesive label, but for this the adhesive used must be chosen carefully so as to stick sufficiently to the fabric without spoiling it or leaving a trace when removed. As a logistics aid the swing tag label is generally one-color and is both machine and human-readable, giving type, size and country of origin. Selling-aid swing tags, frequently affixed through a buttonhole, often display the retailer's logo along with such information as 'pure virgin wool'. They need to be designed for easy removal by the customer. In some countries these labels must also give information on the eco-friendliness (or otherwise) of the garment.

APPAREL LABELING (3) – THE RFID EXPERIENCE

Item-level RFID tagging has not lived up to expectations when it was launched some 15 years ago. An end-user sector where it has flourished is apparel. In this context it is hard not to mention Marks & Spencer, the UK-based retailer generally known as M&S. Apparel accounts for just under half of M&S' revenue, and the group controls its supply lines from the source factory right through to the customer. Since 2014, all M&S clothing items have been RFID-tagged, using UHF (869,5 MHz) technology. Each item is tagged at the point of manufacture, and the tags follow the merchandize until it is sold to a customer. This offers the potential of visibility into the goods' movements throughout the supply chain, into a store's back room and onto the sales floor. It is particularly important in speeding inventory control, which is done using a hand-held reader (Figure 6.4). However, not all apparel items benefit equally from RFID labeling. Typically, the categories that gain the most are items of clothing with many different sizes and colors, with high average selling prices and a long sales life. Experience has shown that for short-term fashion clothing it is preferable to use other forms of label.

Tesco, one of Britain's leading retailers, plans to take smart apparel tagging one step further. In a pilot project, smart robots which roam through the store's clothing department, are doing real-time inventory control; onboard readers perform inventory counts by reading each garment's Gen 2 UHF tag. This robotic system and service is being provided by a US-based technology company which reports that a half-dozen other retailers, globally, are carrying out similar pilot schemes.

Branding and labeling market leaders like Avery RBIS specialize in the apparel and footwear markets, providing both standard apparel labels and also specialties like tactile finishes, foils, silkscreen effects, braille and personalized labels. The importance of smart labeling for the apparel industry was underlined when in early 2016 CCL Label acquired Checkpoint Systems, a leading provider of smart solutions for the apparel sector.

LABELS FOR GARDEN PLANTS AND FLOWERS

Less subject to counterfeiting, flowers and plants nonetheless are a particular challenge for the label converter. The label must be rain and weatherproof over long periods, so paper is rarely used, the main substrates being PE, PET, but also Tyvek or Polyart. Eyelet and loop-lock labels are the most widely used types. For higher unit value items like rose bushes, professionals require attractive, full color labels, often with planting instructions on the back.

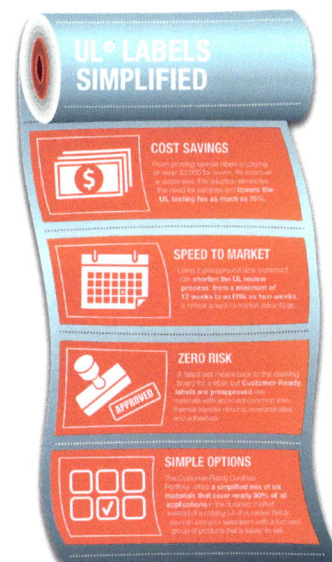

Figure 6.5 Avery's service is one of the ways for label converters to enter the consumer durable sector

The wholesale plant and flower market in Amsterdam is Europe's biggest. Importers and traders need to tag and/or barcode their produce within hours of its arrival, what's more, thanks to the weather, they never know from one day to the next what flowers or plants will arrive. Local label converters have invested in high-volume digital presses and linked-up computer systems so that they can take orders and deliver high-quality, full-color labels all within a matter of hours.

LABELING CONSUMER DURABLES

Durable goods and equipment labels must perform consistently in a variety of hostile environments, conditions and temperatures. For all these categories of product a pressure-sensitive material must be chosen that is cost-effective and durable, which minimizes waste and complies with stringent standards and service requirements. Regulatory requirements are extraordinarily complex in most world markets, depending on the type of appliance, its life cycle and intended use. The EU, for example now requires all electrical white goods (and light bulbs) to carry a label showing the energy efficiency class into which the product falls. Some labelstock suppliers offer a short cut through the regulatory jungle by providing preapproved label solutions. These are designed to give converters a quick, easy, affordable and risk-free way to participate in the high-margin durables market (see Figure 6.5).

Labelstock producers vie with each other in developing specific-use materials for different types of consumer durable. A visit to your local do-it-yourself store will provide any number of examples. UPM Raflatac for example has launched an extra-matte PET label face that meets demand for reliable barcode scanning on durable goods. The matte surface minimizes light-scatter and directs more light back to barcode reader, increasing the effective contrast and the consistency of scanning. Pre-printed information, such as tracking and unit information, is also protected for the lifetime of the label by a topcoat which is scratch, smudge and chemical resistant. Other white goods labels are tested to withstand temperatures up to 200 deg C and to use solvent-free adhesives that nonetheless bond strongly to stainless or galvanized surfaces.

SMART LABELS IN THE HOME AND GARDEN

Certain home appliances (for example, water filters) need regular maintenance or replacement of moving parts. To ensure these deadlines are not forgotten, elapsed time indicator labels can be used. Custom-calibrated for periods up to a year, these labels indicate by a visible sign (generally a color change) when a replacement is due. The labels are 'switched on' by finger pressure. Once activated, a color dye migrates along a display window, showing elapsed time in a range from a few days up to 12 months.

RFID labels in the home are increasingly used as sensors to measure light, temperature or humidity. The information is then fed back to a central control panel which either takes corrective action or warns the householder of an anomaly. This smart system is used for opening/closing shutters, regulating temperature room-by-room, or even deciding when to water the garden!

MEASURING THE RISK

For labeling of other home and garden care durables what is important is the risk associated with the product. Safety instructions labeled on a motor mower must be permanent, and impervious to humidity, oil, abrasions and grass-cuttings. Ladders and stepladders, another high-risk article (in the US, 43 percent of fatal falls in the last decade have involved a ladder) are increasingly being sold with traceable, non-removable labels to enable defective products to be traced back to the manufacturer or importer. Labels for this type of product are mostly one-color, and frequently thermal-transfer printed. Where the manufacturers add their logo or publicity information they may be four-colored.

POINTS TO REMEMBER

- This sector accounts for about one in five of all labels
- Consumer health and safety can be assisted by suitable labels
- Regulations prescribe both ecological and safety stipulations for labels
- A wide variety of label types can be used for household or 'under the counter' products
- Paints, varnishes and sports goods have special labeling requirements
- Luxury goods are frequently protected by sophisticated anti-counterfeit labels, often using RFID
- Digital label printing has opened up new markets where low volumes and fast printing/delivery are needed
- Apparel labeling uses mostly woven and swing-tag labels. RFID labels are increasingly used to speed distribution and stock control
- Strict labeling regulations exist for consumer durables, and special face materials have been developed for labeling these products.

Chapter 7

Industrial labels

The humble industrial label is the unsung hero of the label world. Lacking the hype that goes with primary labels, industrial labels are nonetheless an essential ingredient in keeping the wheels of industry turning, worldwide.

MARKET SIZE AND CHARACTERISTICS

While definitions of an industrial label vary, the MarketsandMarkets Institute this year (2016) estimated at over four percent the annual growth of the world market for industrial labels. This is reinforced by Finat's research, which reckons that in Europe, industrial labels are the biggest end-user sector by volume, and also the most profitable.

In almost all cases, industrial labels must be machine-readable, and be adapted to the industrial process in question. The industrial label is also generally a key component of a track-and-trace system ensuring traceability and combatting the counterfeiter. The different types of industrial end-user are very varied. Some of the main end-user industries are shown in the chart (right), together with the main labeling requirements.

More details on each category are given later in this chapter.

Some regulations are common to all industrial labels. Most apply only to certain categories. Below are some of the rules which label converters must comply with, and which designers, label and pack producers must all be aware of when selecting types faces, substrates, inks, printing processes, label types and methods of application.

REGULATORY BACKGROUND

There are many regulations governing industrial labels, just as there are different categories of risk. Fire

Industry	Main label requirements
Automotive	Security, heat-resistance
Tires	Adhesion to difficult surface
Electronic parts	Temperature-resistance
Printed electronics	High-accuracy, conductive inks
Machinery	Welded or riveted label
Public utilities	High-gloss materials
Defense	Resistance, stringent specs.

Figure 7.1 Hazard labels

extinguishers for example, which may be kept for many years in hot, humid or weather-exposed places, must have a label which retains its color, legibility as well as its resistance to corrosive chemicals over many years. Automotive and tire labels, as described later in this chapter, are also subject to strict labeling requirements. One of the most stringent regulations concerns the labeling of hazardous and non-hazardous materials in drums or barrels. For this there exists now (at last) a single Globally Harmonized System of Classification and Labeling of Chemicals (GHS). This is an internationally agreed-upon system, created by the UN beginning in 1992 and now fully implemented by all major trading countries. It replaces the various classification and labeling standards formerly used in different countries.

The GHS requires manufacturers and exporters of chemicals to provide a label that includes a harmonized signal word, pictogram, and hazard statement for each hazard class and category (see Figure 7.1). It must also state clearly the nature of the hazard and the precautionary measures necessary. Particular care must be taken to identify clearly chemicals which, though inert in themselves, can become volatile if mixed with, or even stored with, other chemical products (failure to respect this precaution is believed to have been the cause of the disastrous explosion at a fertilizer plant in Toulouse in 2001, which cost 31 lives and injured 2,500 people). However, the most hazardous products are drums containing certain chemicals, explosives, petroleum, or radioactive materials. In the USA, labels for these products must comply with complex and sometimes contradictory safety standards established by organizations like the EPA (Environmental Protection Agency), OSHA (Occupational Safety and Health Administration), and the US Department of Trade. For Europe, the EU Directive 2002/95/EC applies. Drum labels in all cases must be permanently applied and must stand up to the harshest of transport and storage abuse. If hazardous chemicals are to be transported by sea, drum labels must now also comply with guidelines such as British Standards BS5609 for weather and salt water immersion. Researchers at one of Avery Dennison's laboratories have developed a labelstock which retains its adhesion

even after three months' immersion in seawater; for this product the company has received a special certification from the globally recognized product security consultancy Underwriters' Laboratory (UL).

Such is the complexity of regulations for durable industrial labels that some labelstock producers offer a service to help converters get fast approval.

SUBSTRATES, ADHESIVES, PRINTING PROCESSES AND INKS FOR INDUSTRIAL LABELS

A) Substrates

Industrial labels are nearly always self-adhesive. Face materials for industrial labels come in many varieties, but heat-resistant filmics predominate. They are most frequently white, corona treated, and with extremely high tear strength in combination with good weather and chemical resistance. They may be overprinted with a clear varnish or overlaminate (usually a clear PET film) to give added protection. Polyester based labeling materials in particular have in recent years come a long way in terms of durability and availability. They now come in high gloss, mat, white and various colors. With the higher temperature ratings of today's polyester labeling materials, a polyester can often be used to lower the cost of labeling in place of some polyimide materials. This substitution is happening particularly in the industrial labels sector.

Metallic foils and metalized substrates are often used for applications such as rating plates, under-the-bonnet, electrical and electronic goods labels.

Whatever the face material used, if the label is destined for outdoor use it needs to be proof against light, and also to withstand a wide ranges of temperatures, water contamination, humidity and abrasion. Inventory control for example is a necessary function in any company, and some fixed assets may have a life of several decades, during which time the identifying label must continue to be legible (and usually machine-readable). Such labels are generally one-color self-adhesives (though sometimes color-codes are used for different classes of asset), and are individually generated using a computer-linked thermal or thermal transfer printer.

Industrial labeling also makes extensive use of thermal transfer (TT) or laser printed labels. These will often be pre-printed with standard information (e.g. a logo, or company name) then just-in-time TT or laser printed with the variable data (which could be a serial or batch number). Sometimes the label converter will do both operations, but generally the variable element is added by the end-user. By pre-printing part of the label the cost per label can be reduced.

B) Adhesives

Many adhesives are used for self-adhesive industrial labels. However rubber based p/s adhesives generally have poor long term aging properties, and for most applications UV acrylic adhesives are preferred, because they give a strong and permanent bond on difficult surfaces, and do not require solvents. Challenging surfaces for industrial adhesives include: glass, metals and polyester. UPM's RX15 is a catalyst activated adhesive for high-temperature automotive labeling, and Wacker has launched a range of silicone polymers for industrial labeling.

C) Printing processes

The durability of pre-printed and non-variable industrial label printing is the most important factor. High performance durable printing using screen process, and hot stamping (using both pigmented and metallic foils), as well as more conventional printing using special inks. Today, industrial labels may use quite sophisticated inkjet printing technology (although this does not always give sufficient durability), as well as laser printing, laser etching and engraving.

D) Inks

Broadly, three families of inks are used for labels: solvent-based, UV curable, and water-based. Solvent inks give good and durable printing results, but are less and less used, as they are not considered to be environmentally friendly. Water-based inks use inks with colorants that are dispersed or dissolved in water. They do not react well to water or humidity and are not suitable for many industrial label applications. UV inks contain pigments, pre-polymers and UV-sensitive materials. They are 'dried' (or cured) by a chemical reaction to UV light. The hardened ink has excellent resistance to water and fading. UV flexo and UV inkjet are the preferred medium for many industrial labels.

Producers of substrates for industrial labeling include, amongst others:

- **Arjobex: 'Polyart Inkjet' one-side coated PET film**
- **Avery Dennison: wide range including labelstock for drum labeling, also flame-retardant PET labelstock for mobile phone batteries, and linerless direct thermal substrates**
- **Hanita: special PET face material**
- **Herma: Labelstock with polypropylene face materials and double-layer adhesive. 'Hermatherm' durable thermal papers**
- **MacTac: certified PET and BOPP substrates**
- **Ritrama: 'Ri-Cote' range with PE facestock – for hazard and drum labeling**
- **Spinnaker Coating: durable polypropylene film with SFA adhesive (developed jointly with ITW Thermal Films)**
- **UPM Raflatac: Full durables range including 'Tyvek' face materials. Designed to meet standards for resistance to heat, chemicals, moisture, mechanical stress and UV light**
- **VPF: PET face material with PE barrier coating**

Manufacturers of thermal and thermal transfer grades include: Hansol, Jujo Thermal, Polyonics and Ricoh.

THE INDUSTRIAL LABEL AS PART OF A LOGISTICS MANAGEMENT SYSTEM

Label converters need to understand at least the broad outlines of their industrial customers' logistics and warehousing system. This is best seen in the

Question: what is a Data Matrix Code?

Answer: a two-dimensional matrix symbology containing dark and light square data modules making up a larger square or rectangular shaped symbol. The Data Matrix code can withstand partial destruction by rubbing or scratching and the encoded data remains readable. The other important feature of the Data Matrix code is that a smartphone can read it and link it to a website or other information source.

Figure 7.2 Sealed aircraft trollies awaiting loading

requirements for medical, automobile and aerospace parts, where security is paramount. Labels will be used a) on the individual part, b) on the case and c) on the pallet. For each of these levels, the labeling requirement will be different but each label must conform to the logistics system used. For industrial labels, even more than for other categories, barcodes, data matrix codes and QR codes are almost universally used. Barcodes have the advantage of being cheap to create and to print. Their main drawbacks are that they do not identify unique items and that scanners need a line of sight to be able to read them. QR and datamatrix codes (see box) are more expensive to create and manage, and can be used to identify single items, but still

need line-of-sight reading. RFID-enabled labels are the most complete (and most expensive) solution, since an RFID-label-based logistics management system can not only determine the number and types of units in every pallet, it can also record date/time of entry/exit from the factory or warehouse. This highly efficient system however requires goods to pass under an RFID-enabled portal each time they are moved to a new location. The costs involved are the main reason why this type of logistics management is not widely used.

SHOCK AND DAMAGE INDICATORS
For the shipping of fragile goods it is important to check at each stage of the transport whether the consignment has been dropped or exposed to seawater. Smart labels containing two or more chemicals are used to give visual warning (for example by a change of color) that an incident has occurred. Such labels are generally single-use, but multiple (re-settable) indicators are also used. Other specialty labels can signal an alert when a product has been tilted. Battery-powered models for all these labels can in addition indicate the time at which the incident occurred. All devices are used in evidence to settle insurance claims.

WHY ARE RFID LABELS NOT USED MORE OFTEN FOR INDUSTRIAL LABELING?
When RFID (radio-frequency identification) labels became generally available in the 1990s many people foresaw their general use in keeping track of any movable asset. Everything from a fleet of hire cars to a batch of hospital trolleys can usefully be tracked and managed via an RFID label. The reason that this has not happened (much) is firstly the cost, secondly the less than 100 percent reliability of the labels, and thirdly the fact that they can only be machine-read at short range. If a hospital trolley gets pushed into the river, no amount of RFID labels will locate it. Even 'active' RFIDs (with battery and transmitter) can only emit a weak signal, and none at all when the battery runs out. However, terrorist attacks have changed people's priorities, and within the past three years airlines have reviewed all their security arrangements: one of the weak points found was the absence of security for

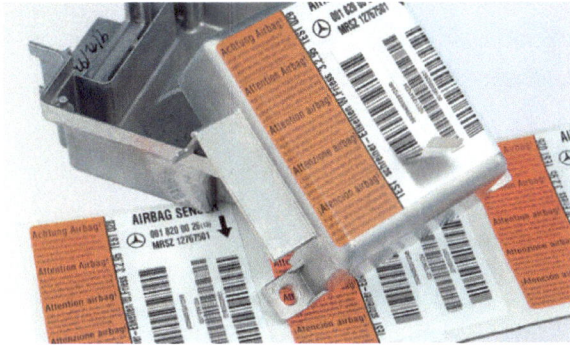

Figure 7.3 Schreiner's security labels offer protection for automotive parts

aircraft galley equipment and trolleys. To remedy this, several RFID-based solutions have been developed. One recent development is called 'Air-Seal'. This 'smart' closing and securing device is based on encrypted banking algorithm technology. When the smart closure is fitted, only authorized staff with the correct ID-tagged security device will be able to open the trolleys. Every action performed on every trolley along the whole supply chain can then be electronically monitored and logged, regardless of location. The system, developed by several European research institutes, is said to be 'tamper-proof, and impossible to duplicate or infiltrate' (see Figure 7.2). At present the system is too costly to install in anything but high-risk locations like aircraft, but similar applications such as hospital equipment could well follow soon.

END-USER APPLICATIONS (1) - AUTOMOTIVE AND AERONAUTICAL LABELING

The average automobile can contain up to 300 labels, some visible, others seen only by your garage mechanic. In most cases they must remain visible and in place over the life of the vehicle, and must resist weathering and extremes of heat and exposure to oils and greases. The same requirements apply to trucks, vans, agricultural tractors and mowers and earth moving equipment, while aircraft will have labels running into thousands covering everything from engine parts to cabin safety notices. Major aircraft manufacturers like Airbus and Boeing have their own labeling standards, often more stringent than the legal

requirements. Similar stringent requirements are required for the labeling of aerospace components.

Automotive parts: the ISO TS 16949 standard stipulates that in serial and spare part production in the automotive industry all elements contained in the final product must be identifiable at all times and uniquely labeled with information such as manufacturing date, expiration date or batch number. In the United States an independent testing institute (Underwriters Laboratories Inc., or UL) inspects products for compliance with consumer protection standards. UL certification is an essential criterion for products supplied to the American automotive industry. Similar regulations exist in Europe and in most other countries. Automotive parts generally carry a secure label, sometimes with RFID chip. The prevalence of counterfeit car parts, and the danger that can result from using substandard parts, means that security labeling (using holograms, safety seals, and so on) on such parts will soon become universal. This applies both to parts for new vehicles and for spares for repair garages.

Labels on new vehicles: each new vehicle, worldwide, has a unique Vehicle Identification Number (VIN), also called chassis number. The VIN label must be securely and permanently fixed (usually to the engine block). It is a favorite target for counterfeiting by unscrupulous used-car dealers. Each label on a new vehicle must conform to rigorous specifications. UV-cured acrylate hot-melt adhesives for example provide especially strong adhesion, even on highly curved, rough, dusty and oily surfaces; face materials must be easily printable, yet scratch and smudge resistant. All this applies to the visible labels, and particularly to under-the-hood labels which need to withstand high temperatures and still remain legible throughout the life of the vehicle.

Substrates and converters: special labelstock developed by Avery Dennison, Raflatac, Herma and other producers can offer the right solution for even the most critical vehicle label application. With all these exacting conditions to be met it is not surprising that just a few label converters specialize in automotive labels. The lucky (or skillful) few are rewarded with lucrative contracts signed with major automobile manufacturers.

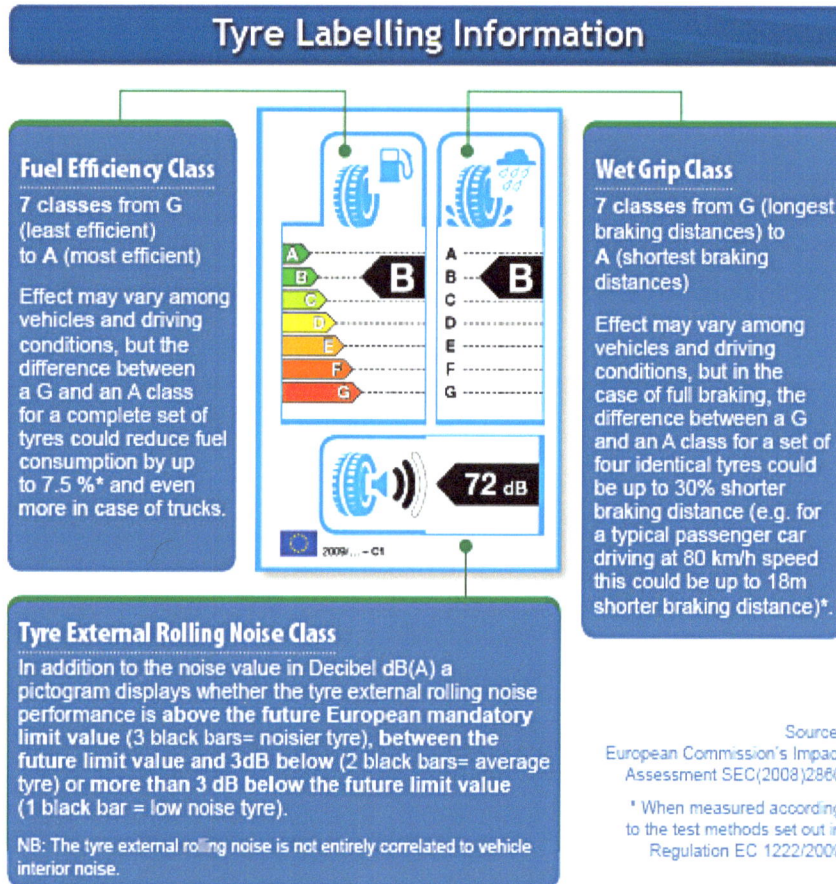

Figure 7.4 EU has strict regulations for tire labeling

Special applications: a section dealing with automotive labels cannot be complete without reference to some advanced electronic labels using OLEDs (Organic Light-Emitting Diodes). OLEDs are made by placing a series of organic thin films between two conductors. When electrical current is applied, a bright light is emitted. They can be made very thin, which makes them suitable for displays and lighting. An early application developed by Schreiner Protech has seen OLEDs used in the interior decoration of certain vehicles. Because they can be made flexible and transparent, they are also being embedded into vehicle windshields.

END-USER APPLICATIONS (2) - TIRE LABELING

Tire labeling is also hemmed in by regulations and with the additional problem that the surface of a tire requires particular adhesives if the label is not to fly off with the first gust of wind. We tend to think of the tire as part of a car, but vans, trucks, aircraft, tractors, and earth moving equipment all need tires, and their specifications are even more stringent than for cars. In Europe, The Tyre Labelling Regulation (EC/1222/2009) which came into force in 2012, introduces labeling requirements with regard to the display of information on the fuel efficiency, wet grip and external rolling noise of tires (Figure 7.4). In the United States the National Highway Traffic Safety Administration has published mandatory regulations requiring tire manufacturers to label their products with ratings on fuel efficiency, traction and durability. It can be said that today, worldwide, tires have become a branded product and the label is how the brand communicates with the consumer. The rough, low-energy surface of a tire requires a rubber-friendly, generous adhesive coating. In some cases backside siliconization is used to stop labels sticking too firmly to the liner. Major labelstock producers have product grades recommended for this application: an example is Germany's VPF which produces a labelstock with glossy white PET face material with permanent lamination of a Polyester barrier, designed for attractive, durable, digital inkjet printing of tire labels (yet another way in which digital printing is making headway). Some converters specialize in the tire label. An example is Serbian label converter Biografika; tire labels supplied to major car manufacturers in ex-Yugoslavia are an important part of this converter's business.

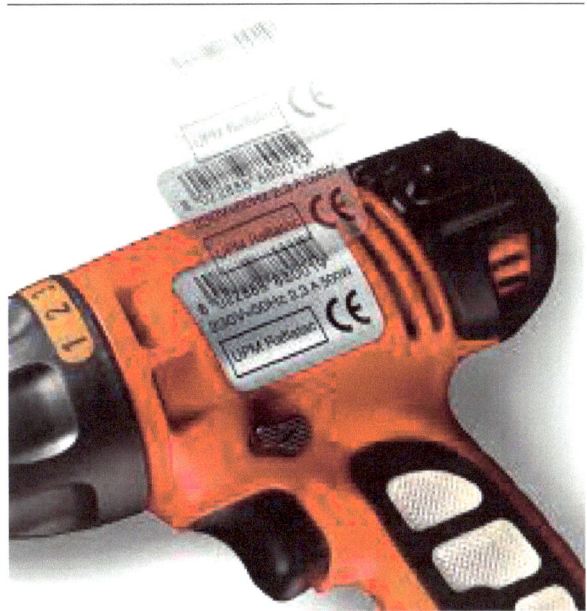

Figure 7.5 Raflatac's special labelstock ensures that electrical goods labels are durable

END-USER APPLICATIONS (3) - LABELS FOR ELECTRONIC AND ELECTRICAL TOOLS/ COMPONENTS

For electronic components a label must perform under some of the most extreme processing and use environments. Labelstock producers offer a range of materials using thermostable face materials (generally polyimides). These are labels which must withstand temperatures up to 200 deg C, exposure to chemicals, humidity and abrasion. They must be clearly machine-readable to ensure efficient track-and-trace. Electronic component labels may be printed using any current print technology, but in practice most use thermal transfer or digital.

All electrical goods must carry warning labels showing power specifications, hazard warning information, wiring instructions, plus a serial number (see Figure 7.5). These labels must remain for the life of the goods, which may be up to ten years or more.

END-USER APPLICATIONS (4) - PRINTED ELECTRONICS

Scientists and engineers have spent many years trying to print microchips and other electronic parts using a roll-to-roll (and generally narrow-web) process. Some people foresaw the day when label converters would partner the chipmakers of the world to produce low-cost components to power the world's computers, smartphones and every other electronic gadget. This has not yet happened, for several reasons, but mainly because today's silicone-based electronics (i.e. practically all of it) does not like being bent or curved. Also, the conductive inks required are expensive (often using silver). The accuracy of printing presses has increased enormously in the past decade, but what is good enough for high-quality primary labels is still not good enough for printing with accuracies in the nanometer (billionth of a meter) range. A company in Accrington, England, reckons it will soon be able to mass-produce electronic devices such as thin-film transistors, using a narrow web process. It is printing sequential coatings onto a filmic or metallic web using electron beams to direct the conductive ink with very high accuracy. The use of graphene-based inks (cheaper than silver) could open the way for a new generation of smart labels with or without printed batteries. The first printed transistors used screen inks that were components for, for example, display electronics. Today's inkjet technology offers wider end-uses. The wiring of circuit boards, for example, can now be printed. This dispenses with the etching/light exposure that is needed for conventional semiconductor manufacturing. Advocates of this process say it saves energy and reduces the need for chemicals, and is therefore an environmentally friendly manufacturing process. The essential advantage of inkjet is its accuracy in jetting minute quantities of conductive inks. Each droplet of a single picoliter (one trillionth of a liter!) can be accurately placed on the substrate at speeds of 6 m/minute. So far this technology is still at the testing stage. It has the potential to open up vast new end-user markets for narrow web inkjet printing.

END-USER APPLICATIONS (5) - LABELS FOR CABLES

Wire and cable labels are an ideal solution for electrical, telecommunication and data communication identification. Wire and cable identification labels are available from many suppliers and in a range of materials suitable for many different environments or applications. Some firms (Brady and Brother are just two examples) specialize in offering a wide range of ID label options, from tags to wraparounds. Portable printers and readers enable the electrician or IT engineer to quickly identify lines during troubleshooting or repair.

END-USER APPLICATIONS (6) - WELDED LABELS FOR MACHINERY OR PARTS

Welded or riveted labels may be used for identifying machinery. They are generally made of plastic, aluminum or steel, and they are permanent. Printing processes for these include laser or conventional engraving / embossing. Potential buyers of used

machinery (or cars) would be well advised to hang onto their money if there are signs that the welded labels have been tampered with!

END-USER APPLICATIONS (7) - MUNICIPALITIES AND PUBLIC UTILITIES

These are perhaps the least glamorous end-users. However, gas, electricity and water suppliers are major users of labels to warn or inform the public. Municipalities and local councils also often have generous budgets for things like anti-litter campaigns.

According to a label converter who specializes in this sector, the labels required use high-gloss polypropylene face materials, and with adhesives that make it 'easy, but not too easy' to remove. Surprisingly, this converter says orders often come with very short lead times. That is why his business has been flourishing since he invested in a digital press.

END-USER APPLICATIONS (8) – MILITARY AND DEFENSE

Governments around the world operate with standardized labeling requirements in all their military and defense procurement specifications. For example, MIL-STD-130 is a US Department of Defense (DoD) mandate that requires suppliers of goods to the DoD to identify items for tracking and efficiency purposes. An added requirement for a Unique Identification Number (UID) may be stipulated in the contract. Each qualifying item must be marked with a permanent 2-dimensional data matrix encoded with the CAGE Code, Serial Number, Part Number and other data elements necessary to construct a Unique Item Identifier (UII). The UID identification must last the expected life of the product or until it comes back to be refurbished.

MIL-STD-130 requires labels to last the expected life of the product even under the harshest environments of water, sand, salt-spray, grease, high temperatures, abrasion, sunlight, chemical exposure, and other elements. Suppliers to the US Department of Defense must be able to prove that their labels meet the requirements of MIL-STD-130, with regard to both Data Matrix Code quality and data accuracy.

For NATO with its many different languages, labels

Figure 7.6 Epson's ColorWorks is a typical hands-free industrial label printer.

make extensive use of a wide range of Military Symbols. These symbols are designed to enhance Nato's joint operability by providing a common standard.

PRINT-AND-APPLY

Print and apply label applicators go from the very simple price-weigh label dispensers, as used by your corner store (and generally using direct thermal printing), to complex industrial applicators incorporating a printer and software to print or overprint labels on demand with such variable data as batch number, price, contents, weight, price and other item-specific information, generally including a barcode.

Print and apply labeling equipment is also used in shipping, warehousing and distribution to provide variable data, for example date of dispatching, which can be used to track and trace goods. Printer manufacturers such as Dymo, Intermec and Zebra have their own proprietary programing languages, and for end users, getting locked into the language of one printer supplier makes it hard to switch brands. End-users in industry or logistics can create their own full color labels on computer, then download them to

hand-held devices. Operators can then add any variable data before printing and applying the finished label. Software built in to the printer will frequently be pre-programed with relevant industry symbols and will ensure compliance with industry standards. The hardware most frequently seen in industrial applications today is the hands-free industrial label printer. Suited to many kinds of trade such as plumbing, electrical installation or maintenance, the hand-held device is ideal for rapidly printing and applying self-adhesive labels for electrics, pipes, cabling, data communications equipment and machinery in offices, homes or construction sites.

The print technology used for print and apply label applicators is typically thermal transfer. The In terms of quantity (but not value) unprinted rolls, most frequently paper, vinyl or polyester are a significant part of total label demand (up to 5 percent according to some experts).

DIRECT PRINTING – A RISK FOR THE INDUSTRIAL LABEL CONVERTER?

The latest inkjet printing technology opens the way for end users to use direct-to-product printing – without the need for labels. In many cases these end-users may have extensive knowledge about inkjet printing already. Will component and product manufacturers take the new direct-to-product inkjet printing technology in-house? And will this make a dent in the demand for industrial labels?

POINTS TO REMEMBER

- National or International regulations govern the labeling of many industrial goods
- Industrial labels must often withstand extreme conditions of temperature and/or humidity, and require special substrates and adhesives
- Labels frequently carry QR codes or RFID chips and are integrated into the end-user's logistics system
- Print-and-apply industrial labels are often produced using a hand-held device
- RFID-based smart labels can provide heightened security
- Many industrial labels are printed using thermal transfer, but screen and hot-foil are also used.

Chapter 8

Special label types and constructions

The purpose of this chapter is not to list exhaustively the hundreds of innovative label constructions used today. It is to draw attention to some of the more interesting and unusual recent developments, and to show the ingenuity with which label converters have solved specific end-user problems, and adapted to public concerns over health, raw materials and the environment.

RE-CLOSABLE LABELS

Back in 1968 two American chemists working with 3M developed an adhesive to make what was almost certainly the world's first re-sealable substrate. Nobody was interested, until twelve years later when 3M launched it as the Post-It - and the rest is history.

More recently, new adhesive formulations have led to the development of the re-closable label, widely used for foods, wet wipes and certain household products.

For foods, the objective is to preserve freshness by providing a better oxygen barrier, and thus to reduce food wastage. Recent developments include special adhesive formulations for closing frozen foods (normal self-adhesive closures do not work at -20 deg C). For wet wipes the re-closable label is designed to keep the moisture in. However wipes contain chemicals, and special adhesives are needed which are not affected by these. Some detergent products are also marketed with a re-sealable label. Recently we have seen pattern-printed cold seal adhesives used as an alternative to self-adhesive complexes.

MULTI-LAYER OR BOOKLET LABELS

Originally developed to meet multi-language regulatory requirements, these labels are now used by FMCG manufacturers to inform the consumer and to present special offers. They can be foldout, fold-in, peel-off or concertina. These features can be used by brand owners to build brand loyalty. Even small containers like lipstick tubes or syringes can now be labeled with extended-text labels that meet compliance requirements for products with active ingredients. Label converters like Schreiner Medipharm are using multi-page RFID-enabled labels (Figure 8.1, multi-page RFID) for hospital use. Converters like CCL Label or J.H.Bertrand have specialty plants in several locations producing booklet and multi-layer labels for food, beverage and pharmaceutical sectors. Equipment suppliers Prati Company, AB Graphic and others have developed ranges of production machinery for making booklet labels. Booklet machines for the pharmaceutical industry or other high security applications may also have inkjet print units to apply barcodes, sell-by dates or sequential numbering to either or both faces of the web.

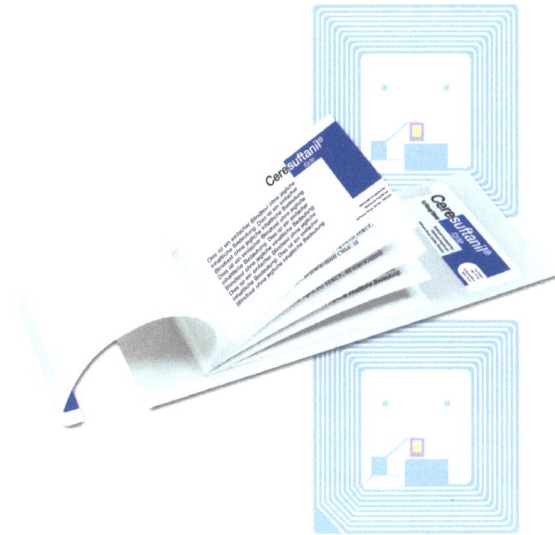

Figure 8.1 Schreiner's RFID-enabled booklet labels enhance medical security

Figure 8.2 Reading a QR code

SMART OR INTELLIGENT LABELS

When this term was first used twenty years ago, the 'smartness' could only be achieved through a microchip in an RFID tag. Some tags were 'active', able to transmit and receive data, most were passive, able to receive and store only. All were too expensive for mass markets, but experts foresaw the day when the ten-cent smart label would dethrone the barcode. That did not happen. The reason is partly that barcodes are by far the cheapest option, and partly

Figure 8.3 The first barcode

that the internet and cloud computing mean that a unique QR (Quick Response) code, datamatrix or serial number can be used to transmit, receive or store 'smartness' held at a location on the web (see Figure 8.2). RFID labels are being used is several end-user sectors (e.g. for registering library books), but the Wal-Mart-style mass market has not materialized.

COMPLEX SECURITY LABELS

Secure closures and 'forgery-proof' labels have been with us for a long time. Developments and improvements are constant because counterfeiters do not stand still. A leading global provider of security labels like Opsec for example uses unique IDs, codes and OVDs (Optically Variable Devices) to provide security, track-and-trace and aesthetic appeal.

For protecting many kinds of assembly line product the company ATT has developed a security label called Seal Vector. ATT describes this as 'a ubiquitous, highly secured and unique code, enabling authentication, identification and serialized traceability of a product or a component'. Decipherable automatically on assembly lines or in the field with regular readers (for example, smart phones), the Seal Vector authenticator protects brand owners, products and consumers. ATT is also possibly the only label-related company to be named a 'Technology Pioneer' by the World Economic Forum in Davos.

Barcodes have both a height and width, although there is normally no interpretive information in the height of the code. However, the height should be sufficient to allow the code to be read efficiently. The dark and light spaces in a barcode represent digitally encoded information, which can be 'read' by devices (barcode readers; barcode scanners), which scan a beam of light across the bars and so pick up signals from the various pulses of reflected light. The dark bars in the code absorb light while the white spaces reflect light, so providing a cost-effective method for automating data collection.

BARCODES

People think of the barcode – if they think of it at all – as the ubiquitous product label that adorns every retail product and keeps supermarket checkout cashiers busy from morning to night. It was the first machine-readable label introduced just 50 years ago (see Figure 8.3), and it revolutionized retailing, warehousing and many other activities. Barcode labels are almost always self-adhesive, and are printed using all usual technologies, including digital. Print precision is essential, and the impression must be scuff and humidity-proof. Various kinds of barcode verifier exist to enable the package or label converter to check the readability of the code. Barcode scanners (or readers) may be fixed (as in most supermarkets) or handheld/wireless as used for warehouse reading. The QR code is a development able to transmit a greater volume of information, and is used frequently for track-and-trace or for promotional marketing purposes. An illegible barcode label causes frustration or worse. It is the last transactional link to the end-user and its importance cannot be overstated.

THIN IS BEAUTIFUL

Wet-glue and in-mold labels have been getting thinner with improved die-cutting and cut-and-stack

Inkjet printing has been around since the 1960s, but the first commercially viable roll-to-roll inkjet printer did not appear until twenty years later. There are various types of inkjet but for today's label industry the so-called piezo technology (also known as drop-on-demand or D-O-D) is used almost exclusively. The principle is simple: the piezoelectric effect is used to force ink through a fine jet. As it falls, the ink forms droplets, and their size can be controlled by the strength of the electric impulse applied. Critical parameters are firstly the distance between jet and substrate and secondly – as also applies with conventional print methods – the absorption characteristics of the face material. The weak point of all D-O-D print heads is that the ink in the heads tends to dry rapidly thus clogging any jet which is not in constant use. To counteract this problem slowly drying printing inks are often used, but this is of course a disadvantage when it comes to drying (or curing) the printed labels.

equipment. Die-cutting both these label types involves cutting right through one or many labels. For self-adhesive labels the problem is more delicate, since the process must cut through the face material and the adhesive without damaging the liner. Progress in self-adhesive die-cutting means that laminates can now be made thinner without the risk of web breaks. All the leading labelstock producers are now offering thinner liners, but beware of web breaks either at the converting or application stages!

Montreal-based ETI Converting recently launched 'Pellicut', a patented technology enabling die-cutting with a regular die down to 0.48 mil (18 microns) polyester film at a speed of up to 750 ft/min. (225 m/min). The company reports that this technology works even with a 12 micron film but they prefer to leave a margin for safety in their claims. Apart from the advantage for the label converter of reduced roll changes, this development helps the converter to gain business with environmentally conscious brand owners and consumers.

THE DIGITAL REVOLUTION

Since the first commercially viable digital label presses became available in the 1990s, three different technologies have been developed. Globally dominating the market is HP Indigo, using a method known as electrophotographic liquid toner. Dry toner is the technology developed by Xeikon. Third and (probably) fastest growing technology is digital inkjet (see box).

HOW DO END USERS BENEFIT FROM DIGITAL LABEL AND PACKAGE PRINTING?

Among the first uses of digital labels were very short run personalized labels for weddings and similar occasions. Major brand owners took a decade or more to realize the potential of personalization as a marketing tool. The chapter on beverages describes examples of successful campaigns promoting soft drinks, beers and similar products.

End-users also benefit from the reduced time to market that is made possible by digital printing (and digital design and finishing). Manufacturers can better adapt to changing market opportunities and so satisfy consumer needs when labels can be designed and printed in a matter of hours without any fixed costs for plates.

The ability to produce short runs at affordable cost means manufacturers of all kinds of product can test consumer reaction by running product trials using a variety of label designs.

POST OFFICES AND PARCEL DELIVERY SERVICES

Most children used to collect postage stamps. Today's stamps no longer hold the same fascination, even for philatelists, but printing and applying them are still big business worldwide. In a little-noticed revolution, self-adhesive postage stamps are rapidly taking over the world. People like you and me appreciate the gain in hygiene of not having to lick

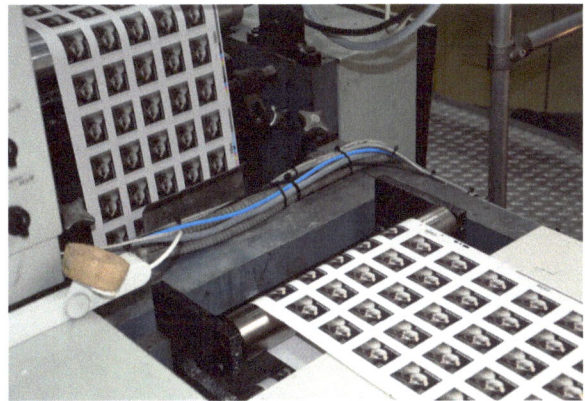

Figure 8.4 Postage stamps being printed on a Nilpeter press

The steamy side of postage stamps

One of the more unusual recent initiatives from Danish label press manufacturer Nilpeter (in cooperation with Avery Dennison and UPM Raflatac) has been the development of special self-adhesive postage stamps. Philatelists – so it seems – dislike self-adhesive postage stamps because they cannot be steamed off envelopes. Now thanks to a new pressure-sensitive laminate with a double layer of two different adhesives, and a combination press supplied by Nilpeter, Post Danmark can print self-adhesive stamps on paper or filmic face materials, safe in the knowledge that philatelists in Denmark and abroad will be able to steam off stamps to their heart's content – another end-user group satisfied thanks to innovative use of self-adhesive technology!

every stamp before using it. Government-owned or controlled printing houses also gain in efficiency by using roll-to-roll presses, usually still with intaglio printing. However, other less secure print technologies are replacing line-engraved intaglio. In some countries, including France, digitally printed postage stamps have made their appearance. This enables cost-conscious official print houses to produce 'special occasion' stamps in limited quantities to mark a sporting event, a centenary or a royal wedding. A novel development in both US and UK is a 'print-your-own-stamps' service. Many businesses apparently prefer to obtain official sanction to print out their own stamps rather than using a franking machine.

ECOLOGY AND ENVIRONMENT

Environmental concerns affect every aspect of the label industry. Substrates, adhesives and inks are all concerned, as are production methods. Brand owners, retail chains and militant consumer associations are asking their label suppliers searching questions. Below are some of the main concerns:

- Do label converters use environmentally friendly materials?
- Do they aim to reduce wasting scarce resources (fuel, water, energy…)?
- Can their products be recycled, if so, how?
- Do they respect regulations on environmental protection and on health and safety laws?

Providing good and reliable answers to these questions must continue to be a major concern for all actors in the label business. Labels, along with other forms of packaging, are not inherently eco-friendly. Consumer protection activists condemn 'excessive packaging' (often forgetting the purpose of the packaging which is to inform, protect and preserve). For many such activists, papermaking destroys forests, plastics pollute oceans, and inks, like adhesives, contain harmful chemicals. There is just enough truth in these allegations for the label industry to address them seriously. Papers along with certain plastics, can be recycled; adhesives do not – except for a few very specific usages – contain toxic substances, and for packages foods, as discussed in Chapter 1, food-grade adhesives have been designed such that even when in direct contact with the product, there is no risk to health. Inks are a problem area. For Europe, REACH legislation has helped develop new generations of ink which do not use potentially dangerous chemicals. In the United States the Environment Protection Agency develops and enforces environmental regulations and, where these are not met, can issue sanctions. Partnerships with industries, businesses, and public organizations work to set goals including conserving water and energy, minimizing greenhouse gases, cutting toxic emissions, re-using solid waste, and controlling indoor pollution.

Consumer pressure (particularly in Germany and UK) has led to the use of low-migration inks for food labels. Finat is a leading force in the promotion and encouragement of the use of environmental management and audit systems like ISO 14001 in the label industry. The Netherlands-based label association has joined forces with its American counterpart TLMI to produce a Guide to Life Cycle Assessment (LCA). Finat environmental consultant Anne Gaasbeek puts it in these words: 'LCA is a widely recognized and scientifically sound method of measuring environmental impact. It takes into account the complete life cycle of a product from the production of the raw materials to the final disposal of the product at the end of its life…It is being used in the self-adhesive industry as a marketing tool to steer product development and develop key performance indicators'. Finat provides guidance documents and case studies to give label converters better understanding of LCA. As Finat General Manager Jules Lejeune has said, brand owners will increasingly come under pressure to show that they and their suppliers are 'green', and label converters in turn will be pressurized to show proof of their ecological credentials. For this, LCA will be an essential tool.

RECYCLING OPTIONS

Throughout the world, label converters and their suppliers are realizing that wasting resources costs them money. Herma is just one labelstock producer who recycles all its waste materials and monitors its use of water and energy. The big stumbling block is liner. Recycling self-adhesive liner is possible, but problematical, due to the presence of residual adhesive and silicone. Several plants exist, including one at Lenzing in Austria, to recycle liner which is then re-used as a raw material for paper production. The limiting factor is transport cost, since end-users must

Finat, TLMI and six other label associations have jointly defined the measures that global label industry associations, suppliers and converters are targeting to improve sustainability. These include:

- **The promotion and encouragement of the use of environmental management and audit systems (ISO 14001, EMAS, LIFE) in the label industry**

- **Enhancing measures to inform, educate and support label producers in meeting current and future label environmental and sustainability targets**

- **Supporting the use of materials and schemes that encourage sustainable and renewable resources, such as FSC, PEFC or SFI**

- **Continued industry development of solutions to maximize cost-effective recovery and recycling of self-adhesive label stock waste**

- **Highlighting the development and use of thinner, lighter label materials**

- **Working towards further reduction in the amount of landfill waste and higher recovery and recycling rates**

- **Having a more prominent industry voice and input into global government, brand owner, packaging and related organizations that are currently impacting on environment and sustainability issues relating to labels and label usage.**

Source: Finat

segregate used liner from their other waste, then wait for the truck which will load it up and take it to the recycling center. Environment consultancy Cycle4Green reckons this service is economic over anything up to 200 km from the recycling center.

While consumers and brand owners clamor publicly for more recycling of packaging, the reality is that when it comes to paying even a few cents extra, the tune changes. Head offices may set out recycling guidelines, but 'front line' packaging plants often ignore them. Ambitious pilot schemes exist to convert liner into pellets for incinerating to produce energy, and they worked well – but only when oil prices were sky-high. So unless there is a technological breakthrough, it looks as if most spent liner will continue to go to landfill.

POINTS TO REMEMBER
- Special label constructions extend the usefulness of the label in all end-user markets
- Digital technology in pre-press, printing and converting offers many new possibilities
- The consumer is increasingly influenced by sustainability, and ecological concerns.
- Recycling, particularly of spent liner, is an unsolved problem for the self-adhesive industry.

Chapter 9

––––––––

The label industry of tomorrow – the forces of change

––––––––

The world of label and package printing for the consumer, retail and industrial sectors, will almost certainly undergo quite significant change over the next five or more years, with an ever-growing list of fundamental drivers of change creating an industry far different to anything seen to-date, and changing at a rate that has never ever been exceeded in the past 500 years.

––––––––

Much of the rapid change impacting on the industry is being driven by things outside of the control of the actual label and package printers themselves, such as globalization, consolidation, e-commerce and internet selling, legislation, changing brand owner requirements, new ways of doing business, cloud computing and the Internet of Things (IoT), counterfeiting and piracy, widespread digitization, consumer trends in shopping and social networking, internet sales, and the rise of own brands.

    The label and package printing industry, to an increasing degree, is only able to react to such changes, rather than driving them. So let's look in a little more detail at these forces of change.

THE FORCES OF CHANGE
Global consumer brands and international super retail groups.

    It is currently estimated that there are probably less than 200 or so global brand and retail super groups that between them will buy, specify or influence close to 70 percent of world label and printed package products. This includes some 15 to 20 global retailers such as Walmart/Asda, Tesco, Sainsbury, Aldi, Lidl, Royal Ahold, Metro; around a dozen global pharmaceutical groups; less than 10 global alcoholic beverage companies; soft drinks giants such as PepsiCo and Coca Cola; the main consumer groups P&G, Unilever, Nestlé, Johnson & Johnson, and others. Between them, these corporations will set the requirements, specifications, standards and levels of service that will influence and drive investment in label materials, technology, products and service worldwide.

    Key forces of change that these bodies are influencing include:
- Demand for consumer (packed and labeled) goods to meet global population growth
- The need for ever increasing multi-language versions and variations
- The need to reduce costs and improve efficiency in the global supply chain
- A requirement to minimize stock holding and move to on-demand printing of labels and packs

The forces of change							
Globalization	**Consumer trends**	**Supply chain efficiency**	**e-commerce**	**Brand protection**	**Environment & sustainability**	**Consumer engagement**	**Product traceabillitty**
More multi-language versions and variations. Use of digital	Own brand ethnic and healthy eating, smaller packs, short runs, digital	Reduce stock holding, on-demand, coding, short lead times	New branding and supply solutions, distribution labels, tracking of goods	Anti-counterfeiting, anti-theft, anti-tamper. Security labels	Thinner materials, linerless, re-use, re-cycling, waste reduction. Regulation	QR codes, Augmented Reality. SnapTags, competitions, games, peel-off, Scratch-off	Bar and other types of codes, RFID, NFC Printed electronics

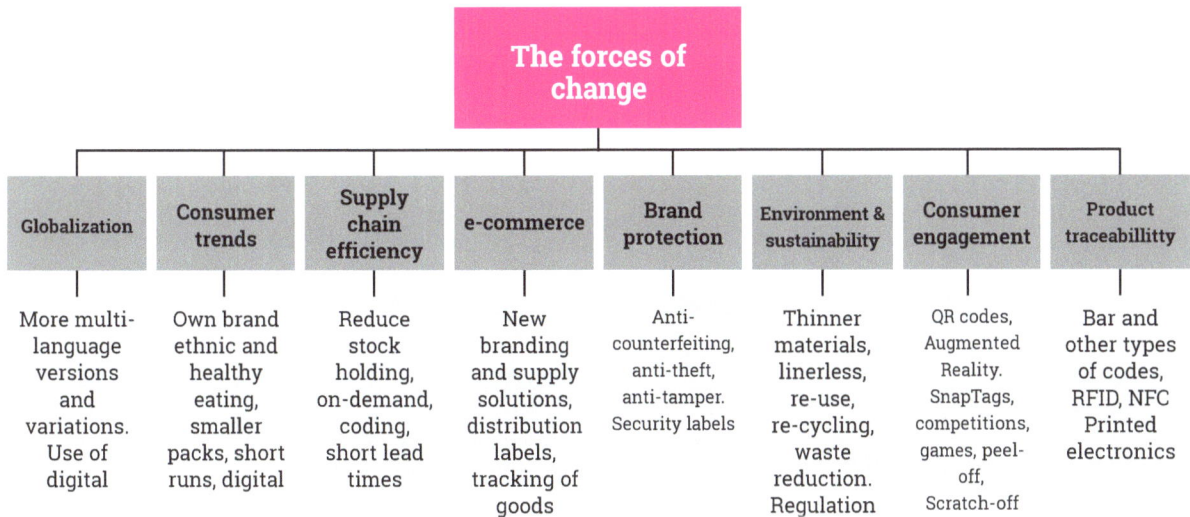

Figure 9.1 Brand owner and end-user forces create change in label materials, production and technology

- Changing trends in packaging, package and label printing, and pack usage – instant foods, microwaveable, re-usable, re-closable, healthy eating, ethnic – and the implications for converters
- Rapid rise in e-commerce shopping. Plus 40 percent in 2014, with 45 percent using mobile phones. Impact on labels and packs
- The need to minimize wastage in fresh and packaged foods – extending food life
- Escalating growth of counterfeit and pirated products – now over 15 percent per annum. 40 percent of goods sold Online are fakes. How do we stop it?
- Major growth in 'Own Brand' packaging
- A requirement to personalize label and packs
- The Internet of Things and how it will increasingly relate to the managing of global label and package printing businesses
- A growing requirement for consumer engagement and social networking through interactive brand labels and packaging
- Ongoing environmental and sustainability challenges and pressures impacting printed label and packaging

- The role of mobile smartphones to communicate with labels and packs
- Consumers in the future to be at the forefront of product authentication and anti-counterfeiting through their mobile phones. Labels and packs will need to have multi-layer authenticity
- Changing international food and product safety regulations and standards – materials, inks, waste, food contact, etc.

Quite simply, there are many forces driving change in the worlds of label and package printing, most of these coming from the major global and national specifiers and influencers, as well as some from national, global and individual country legislation. The label and package printing industry is continually reacting to these forces of change and introducing new technology, materials, production, services and solutions – but largely always playing catch-up. An indication of how change and new label solutions are continually being developed to meet new market demands can be seen in Figure 9.1

Put the forces of change together in a slightly different way and it can quite easily be seen that the role and function of labels (particularly in the developed world markets) has been – and still is -

Label industry growth trends

The role and function of labels in the developed markets is changing

Information carrier	Inactive label	Fixed information	Unsecured data	Minimal food safety	High wastage technology	Unlimited resource

| ↓ | ↓ | ↓ | ↓ | ↓ | ↓ | ↓ |

| Communications medium | Interactive label | Variable information | Secured and traceable data | High food hygiene/ safety | Low wastage technology | Sustainable resource |

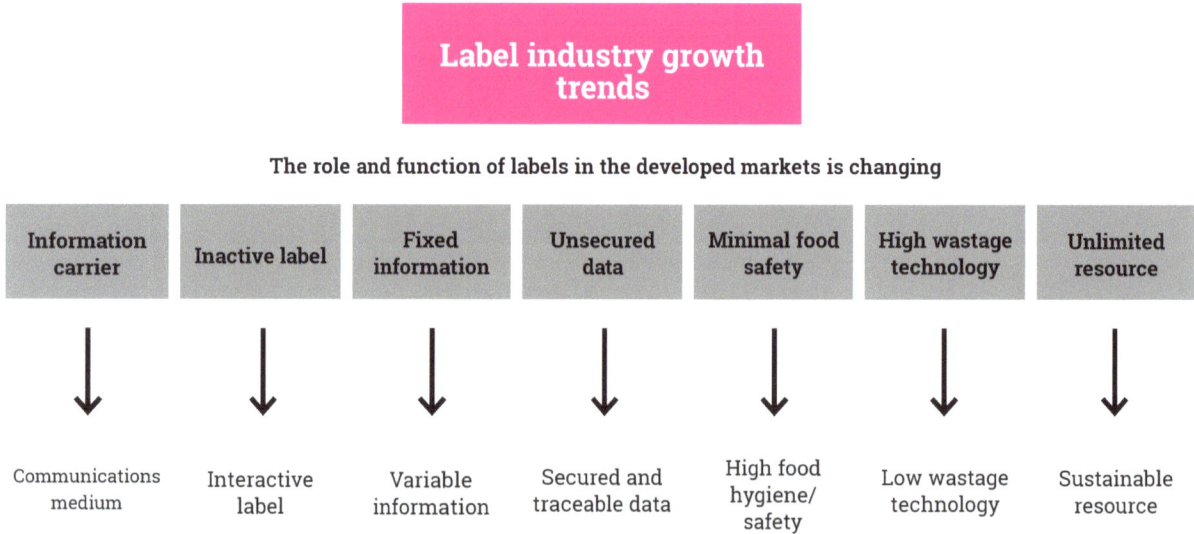

Figure 9.2 Shows how the role and function of labels is changing

undergoing a quite fundamental change (see Figure 9.2) – a change that the label printer and converter must continually adapt to. It should also be noted that digital printing is able to play an increasing role in the changes required by brand owners.

Quite simply, there are many different forces driving change in the worlds of label and package printing, most of these coming from the major global and national specifiers and influencers. The label and package printing industry is continually reacting to these forces of change and introducing new technology, materials, production, services and solutions – but largely always playing catch-up.

Having summarized the key forces of change driving the label industry, it now becomes possible to look in more detail at some of the main areas of products and technology that have come, and are still coming, to the market.

GLOBALIZATION

With global and multi-national brand and retail groups now manufacturing and selling virtually worldwide, it has become necessary for the label industry also to look at new ways of producing and distributing labels to their customers operating in many countries. Do

they follow the brands into new markets and set up new factories? Do they establish partnerships in other countries? Do they undertake mergers and acquisitions? If so, should this be by end-use market sectors, (for example beverage labels, pharma labels, food and supermarket), or by providing new types of services or label solutions?

Some of the world's biggest label converters, like Canadian-based CCL, aim to cover all end-user sectors, worldwide. The CCL Label division has annual sales of 2.4 billion USD, its main strength being in self-adhesive labels, for which it is almost certainly number one worldwide. Recent acquisitions and investments have also made it the leading label converter in China.

Constantia Flexibles (now part of Multi-Colour), which is in the same league, has worldwide sales approaching two billion euros. Constantia's Label Division accounts for 28 percent of the group's sales, and its product range covers self-adhesives, wet-glue, shrink sleeve, in-mold and wraparound. Other label converters, like Barat in France cover just one country and one label type (wines).

Despite some active acquisition activity, the label converting sector remains highly fragmented. There is evidence however, that particularly in the branded goods sector, brand owners are seeking to deal with as few label suppliers as possible to source their requirements worldwide. This trend, if it continues, could hurt the sales of many medium-sized label converters, particularly if they do not have global reach. Small, family-run label converters however need not despair.

Despite the power of the big multi-national brands, most of the world's thousands of label buyers are small companies too, and many of them prefer to buy locally from converters whom they know and trust.

It is nonetheless probable that the coming years will see further concentration of the global label sector as the big players seek further acquisitions to extend their global reach.

The risk associated with this aspect of globalization is that major brand owners will look to local producers in low-cost Asian countries to source their label and packaging requirements. This does not seem to be happening much today, but it is a permanent risk, echoing the shift to Eastern European label suppliers that occurred over the years 1990-2000.

GLOBAL LABEL GROWTH

We have seen that, very broadly, world label markets can be split one third Europe, one third North America, one third rest of the world. We also know that historically, label and packaging markets move roughly in line with Gross Domestic Product (GDP). When economies are expanding, label markets tend to grow at one or two percentage points above GDP. Declining GDP has the opposite effect. Forecast changes in GDP can therefore serve as an indicator of which label markets are likely to be the most vibrant in the near future.

Estimates for selected countries (Figure 9.3) predict modest annual growth in the 0.5 – 2.5 percent range for most of the world's major economies (except China).

Within the Euro zone, only Spain stands out with above three percent projected growth. For Britain and to a lesser extent the rest of Europe, economic growth over the coming years will hang on the final outcome of the Brexit negotiations. The Chinese

Estimated percentage GDP growth 2017	
US	2.5
UK	2.2
Japan	0.0
China	6.2
France	1.3
Germany	1.6
Italy	1.1
World average	3.5

Figure 9.3 Estimated percentage GDP growth 2017. Source: IMF/World Bank

economy, though down from the 8-9 percent growth of recent years, is still slated to continue growing by over 6 percent, and India and Indonesia are also expanding fast. South American markets are characterized by declining GDP in Brazil, Argentina and Venezuela. The volume of world merchandise trade (possibly a better indicator for label markets) has been stuck at around three percent annual growth for the past three years, and here too economists predict that the outlook is not bright.

With these weak estimates for growth in the world economy it is encouraging to note that since 2012, label markets have achieved volume growth well above GDP. A recent Freedonia report estimates future average growth of US label demand at 3.8 percent per year, rising to a value of 19.7 billion USD by 2019. Latest figures from Finat show a buoyant average of over six percent annual growth by volume for European label markets. Per capita self-adhesive

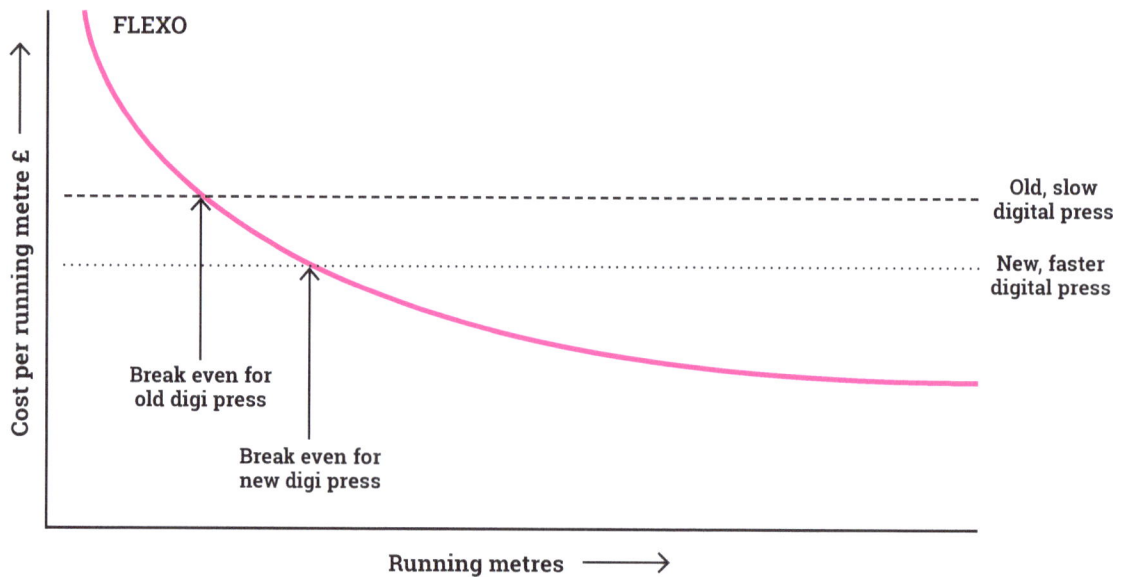

Figure 9.4 As digital technology progresses, the breakeven point with flexo shifts to the right

label consumption in Europe stands at an average of 8.3 square meters per year. Northern Europeans are the biggest users, with Denmark, Ireland, the Netherlands and Britain all-consuming above 14 square meters.

DIGITALLY PRINTED LABELS

That digital printing is revolutionizing the label industry should come as a surprise to no one. For both packaging and labels electrophotographic, dry toner and inkjet printing methods are all suitable. Among the advantages of digital printing are the elimination of traditional printing plates, the cost-effective production of short runs and short delivery times. However, the cost of the print heads and the toner or ink can be high.

For end users two factors are important. Firstly, digital presses are available across the whole price range from less than ten thousand USD to over a million. Secondly, improving digital technology is shifting the breakeven point between digital and flexo (see Figure 9.4) such that in the future, it will be

economic to switch longer runs to digital printing. This trend is already happening: the latest Finat study (for 2015) on run lengths shows conventional run lengths down by a staggering 22 percent (to under 5,000 running meters) compared with 2014. More surprisingly digital print runs are also getting shorter, falling by 12 percent to an average of under 800 running meters.

At a recent meeting which included buyers from Mondelez and Nestlé, the theme was the need to bring products to market even faster. Thanks to digital printing, Nestlé was able to get personalized bottles of Perrier on to supermarket shelves within a month of starting the pilot project. For both these leading brand owners, digital printing will be used more and more both to reduce time-to-market and to boost sales. Consumers, according to Patrick Poitevin of Menendez are willing to pay up to four times as much for a personalized product, and the buzz created on social media gives valuable free advertising to the product.

Figure 9.5 Photo of pilot project for Perrier using HP Indigo technology

THE ENVIRONMENT AND SUSTAINABILITY

The subject of so many heated debates, the ecology and environmental concerns will increasingly affect decision-making in the label industry. These effects will go beyond energy saving and waste-reduction (both of which should increase the converter's bottom line).

In the future we will see more real concern from brand owners and retailers wishing to enhance their ecological credentials to an environmentally responsive public. This may lead to more awareness of recycling, and in particular of release liner recycling; if it does not, the alternative may be legislation – an outcome which label associations such as TLMI and Finat are working to avoid, advocating voluntary incentives.

'Carbon neutral' and 'lowering greenhouse gas emissions' are becoming common terms whenever a new building or factory is planned. Brand owners and retail chains are increasingly looking beyond the sustainability of the labels and packaging they buy, and asking their suppliers to complete detailed questionnaires.

Sustainability goes further than recycling papers and films. New generations of inks, for example, water-based or without VOCs, are environmentally friendly both for the press operator and for the consumer.

Some low migration inks and adhesives are suitable for labeling directly onto foods or medical products. Herma's labelstocks with double-layer adhesive are a significant innovation, combining high tack with low migration. Even washout equipment for presses, anilox rolls and doctor blades are now designed such that only filtered water goes down the drain.

E-COMMERCE - THE 'AMAZON' FACTOR

First it was books and CDs. Now you can buy almost everything on the Internet, and the younger you are, the more likely to use your computer or smartphone to order everything from groceries to ready meals. In some countries as much as 15 percent by value of all retail goods are already being ordered by Internet. The future impact of this growing phenomenon on labels and packaging is not yet clear.

Brand owners quite rightly want their products to stand out on the supermarket shelf. They want to catch the consumers' eye so that they touch, feel and finally buy the product in question. But what if the consumer has only a digital image of the product? The sequence see – touch – buy is broken, and the goods are delivered in a carton which is either neutral, or bears the name of the retailer. It is possible (but unlikely) that in some distant future internet-ordered products will come with distinct, and cheaper, less attractive packaging and labels.

This could happen to products that are ordered, delivered and consumed rapidly, like pizzas, but is unlikely ever to be the case for more long-lasting products. Who wants to get up every morning and see a line of dull and unattractive products on the bathroom shelf?

CONSUMER ENGAGEMENT - THE CONNECTIVITY OF THINGS

Everyone talks about connectivity and it is fast becoming a reality. Homes, cars, refrigerators, even medicines are all becoming part of a joined up world – and labels have an important part to play. The successful Coca-Cola campaign was only the start of a shift in the label industry toward personalized products and interactive promotional marketing. The next step is an app to let consumers scan the lyric on Coke bottle, then record a digital lip-sync video to share on social media.

For many retail products the label will increasingly be the link between the consumer (who wants extra information from the product) and the brand owner (who wants extra information about the consumer). We can even imagine a near future when supermarket labels will flash individual messages ('Hey Mrs. Brown, you visited our website, now try our special offer!') to each customer who walks past.

THREATS AND OPPORTUNITIES IN A CHANGING WORLD

Only a crystal ball gazer would dare to predict which end-user sectors will grow disproportionately, worldwide. We can however draw some tentative predictions, looking at technical/scientific, regulatory and political/macro-economic factors.

TECHNICAL/SCIENTIFIC

Some changes will no doubt be technological. Many people for example believe that linerless self-adhesive labels are about to revolutionize the label business. Others point out that this breakthrough has been imminent for the past twenty years! Recent Labelexpo shows have seen linerless innovations by Ritrama, ETI Converting, Sato, Ravenwood and others. The problem lies, now as in the past, with the applicator. Equipment for applying primary linerless labels is complex (and often expensive!), such that end-users are reluctant to make the change, despite the attraction of a cheaper and more ecological label.

Other future changes could be the development of new materials – nano-particles, for example, or new filmic substrates. Technical developments in digital direct printing are already happening, and this could reduce the demand for labels in certain end-user markets. Another potential threat, paradoxically, is the increasing ease of operation of digital label presses: this has encourages some end-users, including a major French cosmetics group, to bring part of their label requirements in-house.

REGULATORY

In Europe, REACH legislation has been extended (Commission Regulation (EU) 2017/999 of 13 June 2017) to categorize a wider range of substances as toxic. Certain surface treatment directives may also soon be applied to labels. Health and Safety rules governing inks, volatile organic compounds (VOCs) and UV curing are likely to be tightened. In China too, increasing awareness of environmental risks will probably lead to a tightening of regulations, particularly concerning recyclability of packaging materials. However the main risk for the self-adhesive sector is that liner might in the future be reclassified as part of packaging (this is already the case in several European countries including Germany, Austria and the UK). This could put a financial incentive behind liner recycling, and encourage such recent developments as turning used liner into planking, or insulation material for buildings.

POLITICAL/MACRO-ECONOMIC

The lifting of legal barriers (for example the ban on foreign-owned supermarkets in India) will have a dramatic effect on packaging demand in the markets concerned. Events in China, which is now the world's second biggest economy, will affect every aspect of world trade. The lifting of trade barriers and boycotts such as those restricting trade with Russia will help raise demand in many countries. The same benefits will accrue if Iran, a largely closed market of 80 million people, really opens up to international trade over the coming years. Last but by no means least, greater stability and growth in Africa would dramatically increase demand for packaging of all kinds.

All these are potential opportunities for label converters who can identify changing demand and can position themselves at the right place at the right time. However no one should close their eyes to the threats. Another global financial crisis could see label markets go into rapid decline, as they did in 2008/09. Label converters who invested in Ukraine (or Libya, Syria, Iraq...) got their fingers burned, and no-one can say which countries might go the same way in coming years.

DEMOGRAPHIC CHANGE

As some label markets become more technical, others will simply respond to demographics. As middle classes in emerging countries grow in numbers, so will the demand for consumer products, initially with cheaper, more functional labels. In countries like the United States, China and India the richest 5 percent of the population is getting richer, thus boosting the demand for luxury products of all kinds. We also know that two out of every three labels are sold in the developed world, where populations are getting older and living longer. Pharmaceutical and medical labels will benefit. So will the markets for packaged foods, especially those in smaller pack sizes. And a final thought – vanity of vanities – so will all the health and beauty products we need to make us look younger again!

POINTS TO REMEMBER

- Per capita GDP growth is a rough guide to know which label markets are expanding
- Some brand owners want to source their labels globally from just a few converter companies. This will encourage big label converters to expand their global reach
- Supply chain economics and efficiency becoming ever more important
- The breakeven run length between digital and conventional label converting is narrowing
- Tracking and tracing goods and products as they move around the world becoming essential
- Connectivity, and Online shopping, will both change the ways in which labels can be used
- More efficient and economic ways of protecting against counterfeiting, piracy and grey markets will be required
- Environment and sustainability concerns, and in particular the recycling of waste label materials, will be more and more important
- Ageing populations in developed countries will boost demand for certain label categories.
- Need to keep abreast of changing and new legislation relating to labels.

Index

www.ingramcontent.com/pod-product-compliance
Lightning Source LLC
Chambersburg PA
CBHW041723210326

41598CB00007B/758

9781910507131